AF854305

SOLID-STATE POWER SUPPLIES AND CONVERTERS

by

Allan Lytel

Howard W. Sams & Co., Inc.
4300 WEST 62ND ST. INDIANAPOLIS, INDIANA 46268 USA

Copyright © 1965 by Howard W. Sams & Co., Inc.,
Indianapolis, Indiana 46268

FIRST EDITION
ELEVENTH PRINTING—1978

All rights reserved. Reproduction or use, without express permission, of editorial or pictorial content, in any manner, is prohibited. No patent liability is assumed with respect to the use of the information contained herein. While every precaution has been taken in the preparation of this book, the publisher assumes no responsibility for errors or omissions. Neither is any liability assumed for damages resulting from the use of the information contained herein.

International Standard Book Number: 0-672-20310-3
Library of Congress Catalog Card Number: 65-19118

Printed in the United States of America.

PREFACE

The use of semiconductor devices in circuits is growing at a rapid rate. Solid-state power supplies to power these new circuits are also being developed. Unlike the large vacuum-tube power-supply circuits, solid-state circuits are small and compact; this in itself lends to the versatility of the solid-state power supply and converter.

This book is a collection of solid-state power-supply circuits. It is impossible to include all of the existing circuits in one volume. A selection of typical circuits of each of the five different categories has been included, with schematic diagrams, parts values, and all necessary data to construct the circuits. In one reference source this book provides material that would otherwise have to be obtained from the various manufacturers.

Acknowledgment must be given to the following manufacturers, who supplied much of the data from which this book was prepared: Bendix Corp.; CBS Electronics Division, CBS, Inc.; Delco Radio Div., General Motors Corp.; General Electric Co.; International Rectifier Corp.; Motorola Semiconductor Products; Radio Corporation of America; Sarkes Tarzian, Inc.; Texas Instruments, Inc.; Triad Transformer Corp.; and United Transformer Corp.

ALLAN LYTEL

CONTENTS

SECTION 4

SECTION 5

INTRODUCTION

This book covers a wide range of solid-state rectifiers and power-supply circuits as used in industrial equipment, hobby circuits, and transmitter-receiver units of all types. Each circuit is complete with parts values and all the information necessary to build the power supply.

CIRCUIT WIRING

Although power transistors and rectifiers are very rugged devices that can undergo considerable rough treatment, unnecessary mechanical shock is to be avoided. It is important to avoid extreme shock or excessive bending or twisting of the device leads when they are being soldered. When operated within their normal ratings, these components will provide a long and useful life. It is possible, however, that even small excessive voltage or excessive power demands may instantaneously destroy a unit such as a power transistor.

When you are building or experimenting with new circuits, it is important to take certain precautions. The polarity of the power-supply voltage for the circuit should be checked and then double checked. An incorrect polarity can damage the transistors as well as any electrolytic capacitors used in the circuit. After a new circuit is built and it is being tested, the voltage should be applied in easy stages, beginning with a value below normal, to check the basic wiring to see if the circuit operation is correct. It is important to be very careful of high-voltage surges at all times. If an unregulated 12-volt electronic power supply is constructed, it is possible to load this power supply with a storage battery in order to stabilize the voltage.

If the power-supply circuit appears to be operating properly, the rectifier current and collector current in the power-transistor stages of a regulator, where one is used, should be measured and the bias adjusted if necessary. This will prevent any long-term damage by operating a transistor beyond its normal ratings.

All of these circuits have been tested; they have been conservatively designed and will provide excellent performance. No layouts and mechanical details are given since they vary widely with different requirements as well as with the sizes and shapes of the components employed. The choice of component parts is very important since the performance of these circuits depends on the quality of the components and the care taken in layout as well as the care taken in construction. In almost all cases, transformers can be obtained from local parts dealers by specifying the required characteristics. The wattage ratings given for resistors are based on methods of construction that provide adequate ventilation, and compact installations with ventilation less than normal may require resistors of higher-wattage ratings.

Capacitor voltage ratings are the minimum d-c working voltages required. Paper, mica, or ceramic capacitors with higher voltage ratings than those given may be used if the physical sizes of such capacitors will not affect the equipment layout. Where electrolytic capacitors with much higher voltage ratings than those given are used, they may not "form" completely at the circuit operating voltage, and as a result, the effective capacitances may be below their rated values.

For reference, some major manufacturers of power transformers follow:

Columbus Process Company
Columbus, Indiana

Microtran Company, Inc.
145 East Mineola Avenue
Valley Stream, New York

Mid-West Coil and Transformer Co.
1642 North Halstead
Chicago, Illinois

Stancor Electronics, Inc.
3501 West Addison Street
Chicago, Illinois

Triad Transformer Corporation
4055 Redwood Avenue
Venice, California

Thompson-Ramo-Wooldridge, Inc.
Electronic Components Division
666 Garland Place
Des Plaines, Illinois

Thordarson
7th and Bellmont
Mt. Carmel, Illinois

United Transformer Corporation
150 Varick Street
New York, New York

HEAT SINKS

Temperature extremes can be damaging to a transistor. Power transistors should usually be mounted on a heat sink or radiator, and where the collector is above ground potential, the sink should be insulated from the chassis. It is possible to bolt the transistor to the chassis with insulated bolts and to use a washer between the transistor and chassis.

Excessive heat is an enemy of power transistors. Since heat increases the collector cutoff current, which reduces the power output, further heat is then developed; it is possible that this might result in thermal runaway. A circuit can be stabilized by using a thermistor or temperature-sensitive resistor in the base circuit so that an increase in temperature decreases the base-to-emitter voltage, thus stabilizing the collector cutoff current.

Transistor heat sinks are necessary in most power-transistor applications. Fig. 1 shows the Delco 7281351 heat sink made of extruded aluminum. This has a nominal weight of 8 ounces and a surface area of about 100 square inches. The mounting of two different types of transistors in this heat sink is shown in Fig. 2. The use of these mounting kits will allow the insulated mounting of transistors to this heat sink. The effectiveness of a particular heat sink depends

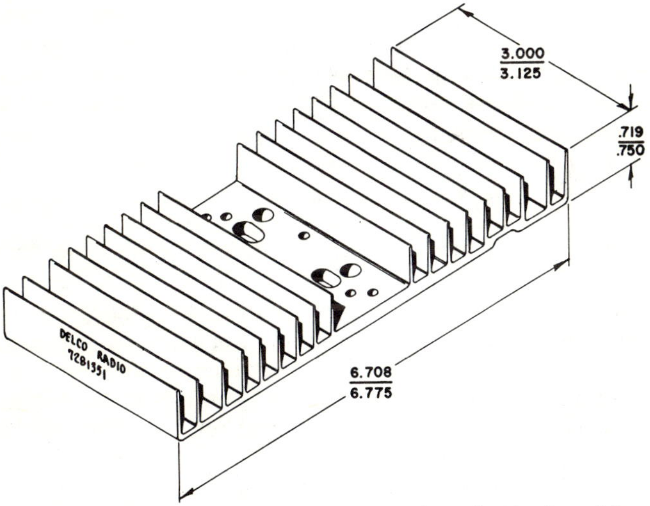

Courtesy Delco Radio Div., General Motors Corp.

Fig. 1. Delco 7281351 heat sink.

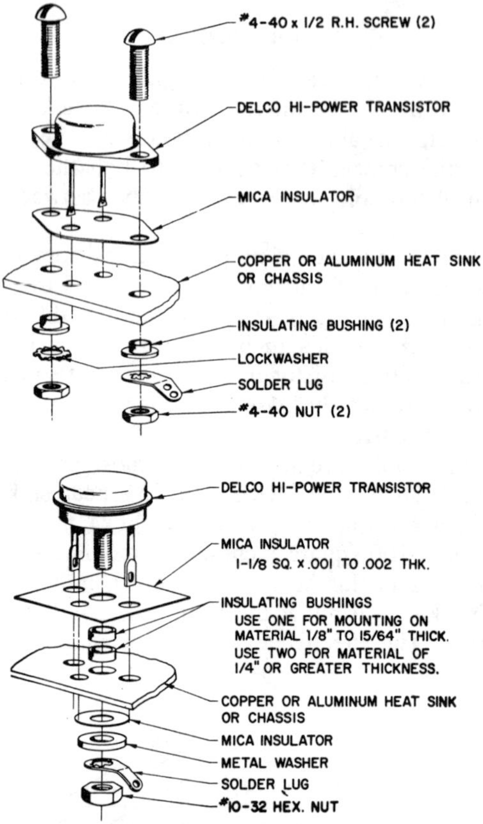

Courtesy Delco Radio Div., General Motors Corp.

Fig. 2. Mounting transistors on heat sink.

on the contact interface between the transistor and the heat-sink surface. It is necessary that the most intimate contact be made for heat transfer, and the effectiveness of this contact depends on the meeting of the surfaces and the pressure holding them together. The use of a grease or oil is a beneficial technique to minimize the effect of any surface irregularity where the two surfaces meet. A suggested type is silicone oil type 200 (Dow Corning Corp., Midland, Michigan).

A different type of heat sink (Delco type 7281357) is shown in Fig. 3. This type of heat sink has an insulated spacer that is used to allow insulation of the entire heat sink from the chassis so that the transistor can be mounted directly to the heat sink.

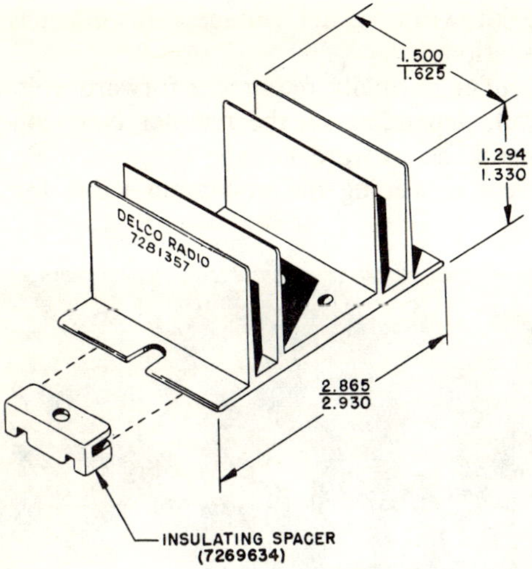

Courtesy Delco Radio Div., General Motors Corp.

Fig. 3. Delco 7281357 heat sink.

SILICON RECTIFIERS

Silicon rectifiers can be considered as power diodes using semiconductor materials. They have a low forward resistance and a high back resistance; these rectifiers can be operated at temperatures up to 200°C, and with currents of 100 amperes or more at voltage levels up to 1000 volts.

There are many sizes and shapes of silicon rectifiers, including the smallest types usually used for power supplies (Fig. 4); these glass diodes have pigtail leads. Plug-in types (Fig. 5) have higher ratings than the glass diodes. The largest solid-state rectifiers are shown in Fig. 6. For rectifier stacks, units such as those shown in Fig. 7 are used.

Silicon rectifiers provide a high-rectification efficiency because of their high forward-to-reverse current ratios; when they are properly used, they are excellent rectifiers in power-supply circuits. Although they can operate at high temperatures, silicon rectifiers are sensitive to sudden temperature change; thus, any sudden rise in junction temperature can cause a rectifier failure.

During reverse bias a small amount of current flows, but at a specific reverse voltage, depending on the type of diode, there is a very sharp increase in reverse current. This voltage is known as the avalanche, breakdown, or zener voltage. This effect is used in zener diodes for regulation.

A silicon rectifier normally requires a forward voltage of between 0.4 and 0.7 volt, depending on the rectifier type. Since the rectifier has a small mass, the forward voltage drop must be carefully controlled to prevent exceeding the maximum value for the particular device.

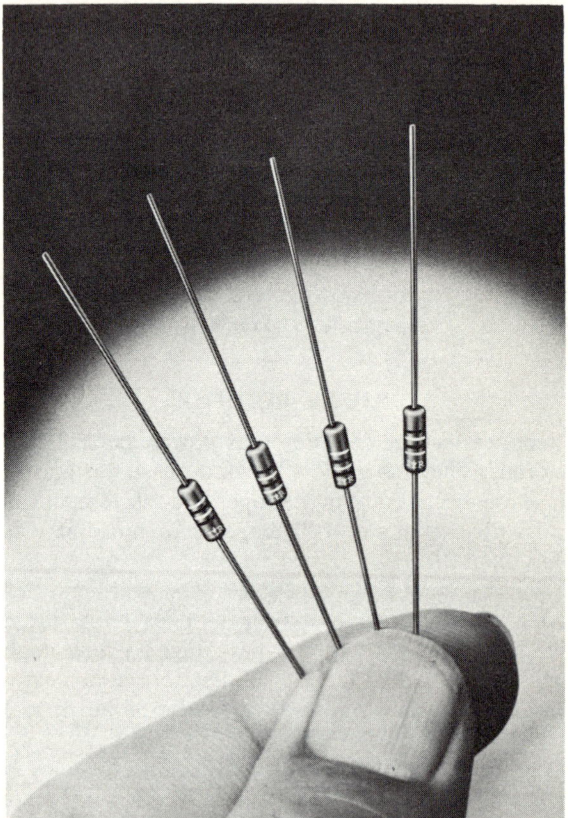

Courtesy International Rectifier Corp.

Fig. 4. Glass-diode silicon rectifiers.

Silicon rectifiers were originally developed for use in dc-to-dc converters, battery chargers, mobile power supplies, and other applications. Because of their excellent characteristics, they are now being used in power supplies for all types of electronic equipment.

In addition to the usual diode rectifier, the silicon controlled rectifier (SCR) is a powerful circuit aid where variations in power are required. Control of the gate input makes provisions for different firing points on the a-c cycle. SCR's are generally used in power-control circuits, while standard power-supply design most often uses diodes.

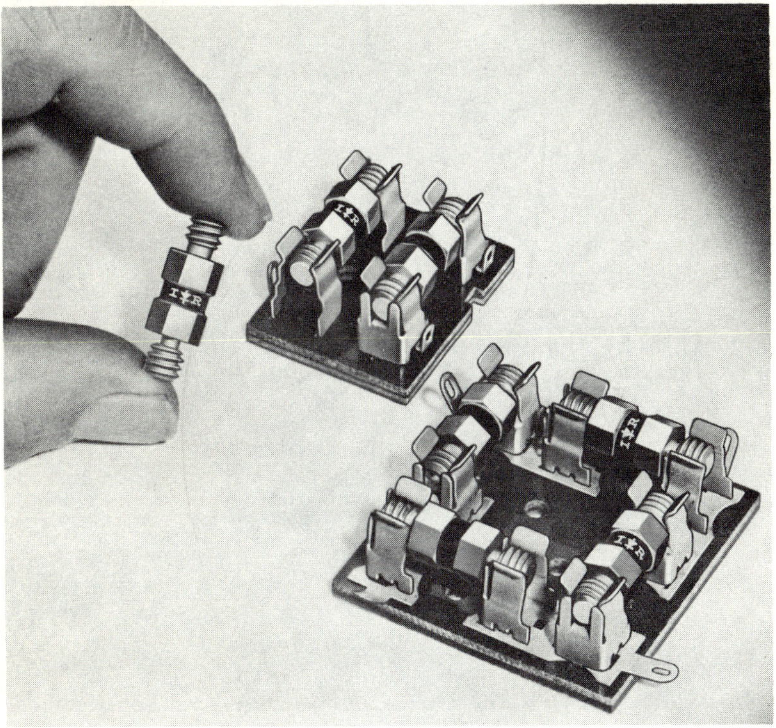

Courtesy International Rectifier Corp.

Fig. 5. Plug-in silicon rectifiers.

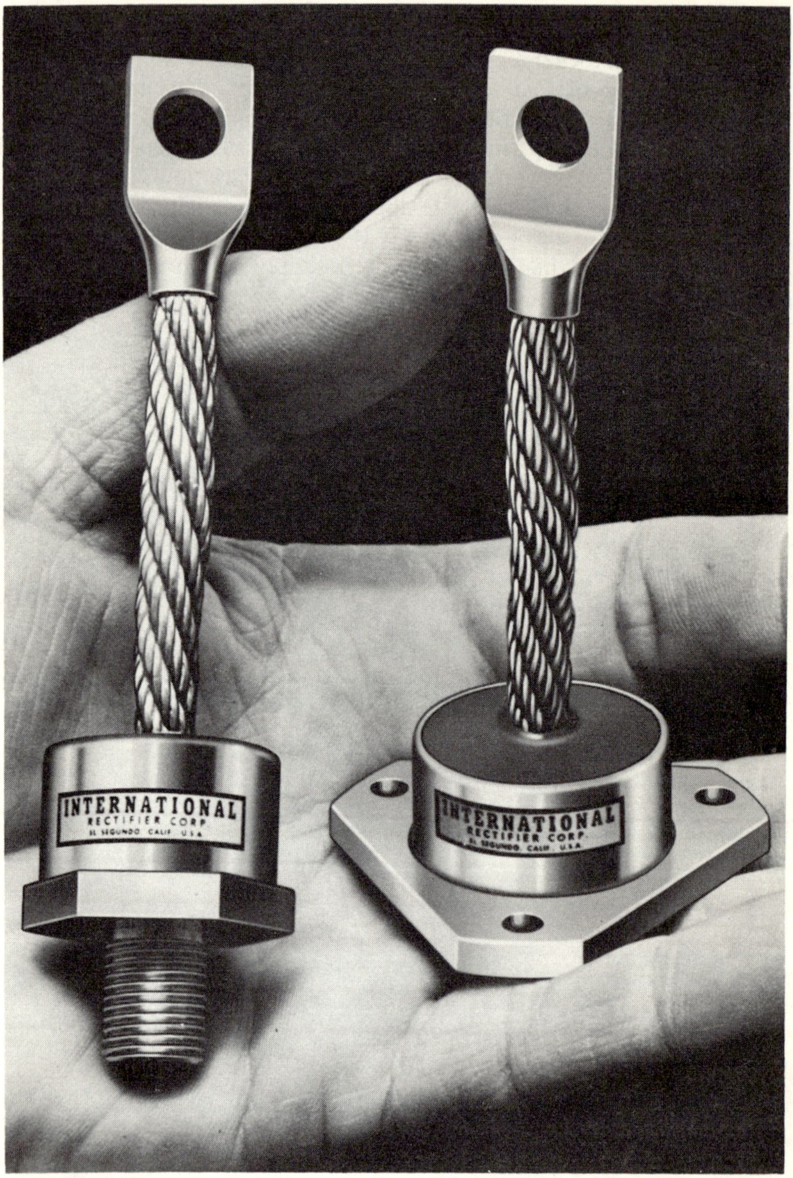

Courtesy International Rectifier Corp.

Fig. 6. High-ampere solid-state rectifiers.

Courtesy International Rectifier Corp.

Fig. 7. High-power silicon-rectifier stacks.

RECTIFIER CHARACTERISTICS

In choosing a rectifier circuit the parameters under consideration are the d-c voltage required, the current required, and the amount of ripple in the output circuit. Ratings for silicon rectifiers are important in considering their specific applications in a particular circuit. The *maximum peak-reverse voltage* (inverse voltage) is the highest amount of reverse voltage that can be applied to a particular rectifier before avalanche breakdown is reached. These ratings range from about 50 volts up to as high as 1000 volts for diodes with a single junction. It is possible, of course, to have any number of rectifiers in series to provide the necessary overall peak-reverse voltage required by a particular circuit.

Chart 1. Characteristics of

				SINGLE PHASE HALF WAVE (1-1-1-H)	SINGLE PHASE CENTER-TAP (2-1-1-C)	SINGLE PHASE BRIDGE (4-1-1-B)
TYPE OF CIRCUIT						
PRIMARY / SECONDARY				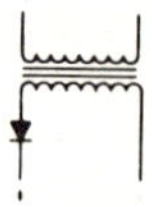	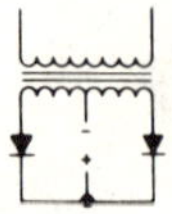	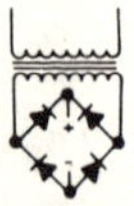
ONE CYCLE WAVE OF RECTIFIER OUTPUT VOLTAGE (NO OVERLAP)				0 	0 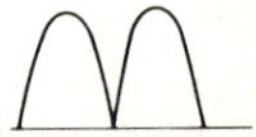	0 
NUMBER OF RECTIFIER ELEMENTS IN CIRCUIT				1	2	4
RMS D.C. VOLTS OUTPUT			=	1.57	1.11	1.11
PEAK D.C. VOLTS OUTPUT			=	3.14	1.57	1.57
			=	3.14	3.14	1.57
PEAK REVERSE VOLTS PER RECTIFIER ELEMENT			=	1.41	2.82	1.41
			=	1.41	1.41	1.41
AVERAGE D.C. OUTPUT CURRENT			=	1.00	1.00	1.00
AVERAGE D.C. OUTPUT CURRENT PER RECTIFIER ELEMENT			=	1.00	0.500	0.500
RMS CURRENT PER RECTIFIER ELEMENT		RESISTIVE LOAD	=	1.57	0.785	0.785
		INDUCTIVE LOAD	=	----	0.707	0.707
PEAK CURRENT PER RECTIFIER ELEMENT		RESISTIVE LOAD	=	3.14	1.57	1.57
		INDUCTIVE LOAD	=	----	1.00	1.00
RATIO: PEAK TO AVERAGE CURRENT PER ELEMENT		RESISTIVE LOAD		3.14	3.14	3.14
		INDUCTIVE LOAD		----	2.00	2.00
% RIPPLE (RMS OF RIPPLE / AVERAGE OUTPUT VOLTAGE)				121%	48%	48%
				Resistive Load		**Inductive**
TRANSFORMER SECONDARY RMS VOLTS PER LEG			=	2.22	1.11 (TO CENTER-TAP)	1.11 (TOTAL)
TRANSFORMER SECONDARY RMS VOLTS LINE-TO-LINE			=	2.22	2.22	1.11
SECONDARY LINE CURRENT			=	1.57	0.707	1.00
TRANSFORMER SECONDARY VOLT-AMPERES PER LEG			=	3.49	1.57	1.11
TRANSFORMER PRIMARY RMS AMPERES PER LEG			=	1.57	1.00	1.00
TRANSFORMER PRIMARY VOLT-AMPERES PER LEG			=	3.49	1.11	1.11
AVERAGE OF PRIMARY AND SECONDARY VOLT-AMPERES			=	3.49	1.34	1.11
PRIMARY LINE CURRENT			=	1.57	1.00	1.00
LINE POWER FACTOR				----	0.900	0.900

common rectifier circuits.

THREE PHASE STAR (WYE) (3-1-1-Y)	THREE PHASE BRIDGE (6-1-1-B)	SIX PHASE STAR (THREE PHASE DIAMETRIC) (6-1-1-S)	THREE PHASE DOUBLE, WYE WITH INTERPHASE TRANSFORMER (6-1-1-Y)	*Assumes zero forward drop and zero reverse current in rectifiers and no AC line or source reactance. TO DETERMINE ACTUAL VALUE OF PARAMETER IN FIRST COLUMN, MULTIPLY FACTOR SHOWN BY VALUE OF:
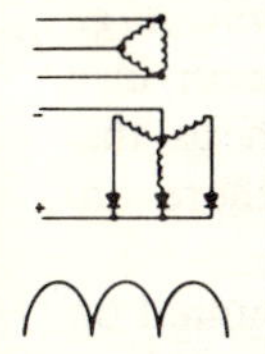	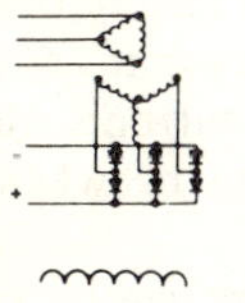	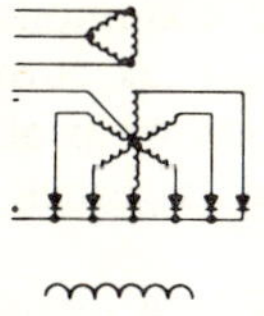	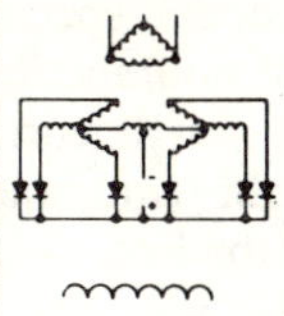	
3	6	6	6	
1.02	1.00	1.00	1.00	X AVERAGE D.C. VOLTAGE OUTPUT
1.21	1.05	1.05	1.05	X AVERAGE D.C. VOLTAGE OUTPUT
2.09	1.05	2.09	2.42	X AVERAGE D.C. VOLTAGE OUTPUT
2.45	2.45	2.83	2.83	X RMS SECONDARY VOLTS PER TRANSFORMER LEG
1.41	1.41	1.41	1.41 (DIAMETRIC)	X RMS SECONDARY VOLTS LINE-TO-LINE
1.00	1.00	1.00	1.00	X AVERAGE D.C. OUTPUT CURRENT
0.333	0.333	0.167	0.167	X AVERAGE D.C. OUTPUT CURRENT
0.587	0.579	0.409	0.293	X AVERAGE D.C. OUTPUT CURRENT
0.578	0.578	0.408	0.289	X AVERAGE D.C. OUTPUT CURRENT
1.21	1.05	1.05	0.525	X AVERAGE D.C. OUTPUT CURRENT
1.00	1.00	1.00	0.500	X AVERAGE D.C. OUTPUT CURRENT
3.63	3.15	6.30	3.15	
3.00	3.00	6.00	3.00	
18.3%	4.2%	4.2%	4.2%	
Load or Large Choke Input Filter				
0.855 (TO NEUTRAL)	0.428 (TO NEUTRAL)	0.740 (TO NEUTRAL)	0.855 (TO NEUTRAL)	X AVERAGE D.C. VOLTAGE OUTPUT
1.48	0.740	1.48 (MAX.)	1.71 (MAX.—NO LOAD)	X AVERAGE D.C. VOLTAGE OUTPUT
0.578	0.816	0.408	0.289	X AVERAGE D.C. OUTPUT CURRENT
1.48	1.05	1.81	1.48	X D.C. WATTS OUTPUT
0.471	0.816	0.577	0.408	X AVERAGE D.C. OUTPUT CURRENT
1.21	1.05	1.28	1.05	X D.C. WATTS OUTPUT
1.35	1.05	1.55	1.26	X D.C. WATTS OUTPUT
0.817	1.41	0.817	0.707	X AVE. LOAD CURRENT X SEC. LEG VOLT. / PRI. LINE VOLT.
0.826	0.955	0.955	0.955	

Courtesy General Electric Co.

Current is usually specified in three different ways. The *maximum average forward current* is that rectifier current which is allowed to flow in the forward direction at a specific ambient or case temperature. Average currents are typically from 0.5 ampere to as much as 250 amperes for special rectifier types. The *maximum recurrent forward current* is the maximum repetitive instantaneous current in a forward direction under specific conditions. The *maximum surge current* is the greatest nonrepetitive peak current that can occur in a single forward cycle. Surge currents or peak currents such as this usually occur when the equipment is turned on or when there are unusual transients in the line.

To prevent excessive heat rise, silicon rectifiers can be mounted on heat sinks attached to the heat-conducting side of the rectifier. A heat sink will allow the rectifier to dissipate a large amount of heat, thus protecting the rectifier circuit against damage. The size of a heat sink for a specific application depends on the maximum average forward current and the allowable rectifier temperature.

It is possible to use silicon rectifiers in series when the expected applied reverse voltage is greater than the maximum peak-reverse voltage rating for a particular rectifier cell. When cells are used in series, it is very important that the applied voltage be divided equally across the individual rectifiers; otherwise an instantaneous voltage greater than the rated maximum might be applied across an individual cell. For this purpose shunt resistors and shunt capacitors are used to equalize the voltage applied to the series resistors; both resistors and capacitors should be used in a circuit if it is going to be carrying both dc and ac.

Rectifiers can be used in parallel when the maximum average forward current required for a particular application is greater than the maximum average forward current rating for an individual rectifier cell. A resistor or inductor should be added in series with each cell in order to avoid differences in voltage from appearing across the parallel rectifiers.

Characteristics of seven basic rectifier circuits are shown in Chart 1. Assuming a zero forward-voltage drop and zero reverse current in the rectifiers, the table compares these seven circuits for various parameters. There is a small error in using this table, but this is usually not significant because the voltage drop in the forward direction is assumed to be zero, as is the reverse current.

The single-phase half-wave circuit (A) delivers one current pulse for each cycle of a-c voltage input. The single rectifier cell has the entire current flow, and there is a great deal of output ripple. A circuit of this type is used principally in low-voltage high-current applications, as well as in low-current high-voltage applications.

The single-phase full-wave circuit (B) has a center-tapped high-voltage winding, with a higher peak-to-average voltage ratio and less ripple than the single-phase half-wave circuit. Large audio amplifiers and television receivers quite commonly use this type of rectifier.

Four rectifiers are used in the single-phase full-wave bridge circuit (C) which does not require the use of a transformer center tap. This circuit provides twice as much output voltage as the single-phase full-wave circuit for the same transformer voltage. The single-phase full-wave bridge applies half as much peak-reverse voltage to each individual circuit cell, and allows only half of the total current to flow through each cell. Circuits of this type are quite often used in radio transmitters.

Three-phase circuits are usually found in industrial equipment and in transmitters of high power. The three-phase wye circuit (D) uses three rectifier cells and has considerably less ripple than any of the single-phase circuits. This circuit also allows one-third of the total current to flow through each individual rectifier cell; a circuit of this type is used in alternator rectifiers in automobiles.

The three-phase wye, full-wave bridge circuit (E) uses six rectifier cells. In this circuit two half-wave rectifiers are connected in series across each leg of a high-voltage transformer; this circuit delivers twice as much voltage output as the three-phase wye, half-wave circuit for the same voltage conditions. There is also an extremely low ripple percentage and a very low ratio of peak-to-average voltage.

The six-phase star circuit (F) uses six rectifiers and allows the least amount of total current to flow through each cell. Only one-sixth of the total current will flow through each rectifier cell.

The three-phase double wys, an interphase-transformer circuit (G), uses six individual half-wave rectifiers in parallel and delivers six current pulses per cycle. This circuit will deliver twice as much output current as the three-phase wye, half-wave circuit.

Filter circuits are used to smooth the ripple in the output of a rectifier circuit, and a smoothing filter uses an iron-core choke together with one or more capacitors. In choosing a filter for a particular circuit, the load impedance is an integral part of the filter design since the load affects the performance of the filter. The purpose of the choke is to provide a high impedance to the ripple voltage, while the capacitor stores energy between voltage peaks.

Charts 2, 3, and 4 are rectifier selection charts. Complete instructions on how to use these charts precede them.

A HANDY SELECTION CHART TO SILICON AND GERMANIUM RECTIFIER COMPONENTS

1. Select applicable chart.
2. Locate average current I per rectifier cell on horizontal axis.
3. Move vertically to intersection of average current requirement with recurrent peak reverse voltage (PRV) requirement.*
4. Find the first numbered dot above and to the right of the intersection point located in Step 3. The corresponding rectifier cell type indexed in the box below the dot is the optimum rated cell for the application.

ADDITIONAL INFORMATION:

1

2. Ambient = 55°C Free Convection (Rectifier cell furnished with own heatsink).
3. Ambient = 125°C Free Convection (Rectifier cell furnished with own heatsink).
4. Stud mounted rectifiers (cell must be attached to heatsink provided by user). Stud Temperature = 150°C.

2

Rectifier Circuit	Average Current Per Cell
Single Phase Half Wave	I = DC Load Current
Single Phase Centertap	I = ½ x DC Load Current
Single Phase Bridge	I = ½ x DC Load Current
Three Phase Halfwave	I = ⅓ x DC Load Current
Three Phase Bridge	I = ⅓ x DC Load Current

3

Rectifier Circuit	Recurrent Peak Reverse Voltage
Single phase halfwave —resistive load	PRV = 1.41 x AC supply voltage
—capacitive load	PRV = 2.82 x AC supply voltage
Single phase centertap	PRV = 2.82 x AC secondary voltage (line to centertap)
Single Phase bridge	PRV = 1.41 x AC supply voltage
Three phase halfwave	PRV = 2.45 x AC secondary voltage (line to neutral)
Three phase bridge	PRV = 2.45 x AC supply voltage (line to line)

*Additional PRV rating will be necessary if voltage transients above the normal line voltage occur.

4

Example:

Requirements: 5 amps, 350 volts PRV per rectifier cell, 55°C free convection ambient. Cell with heatsink required.

Solution: In Chart **2**, point 3 is just above and to the right of intersection of 5 amperes and 350 volts. In corresponding box below chart, point 3 is identified as General Electric type 4JA2011D rectifier stack.

Chart 2. Rectifier selection chart.

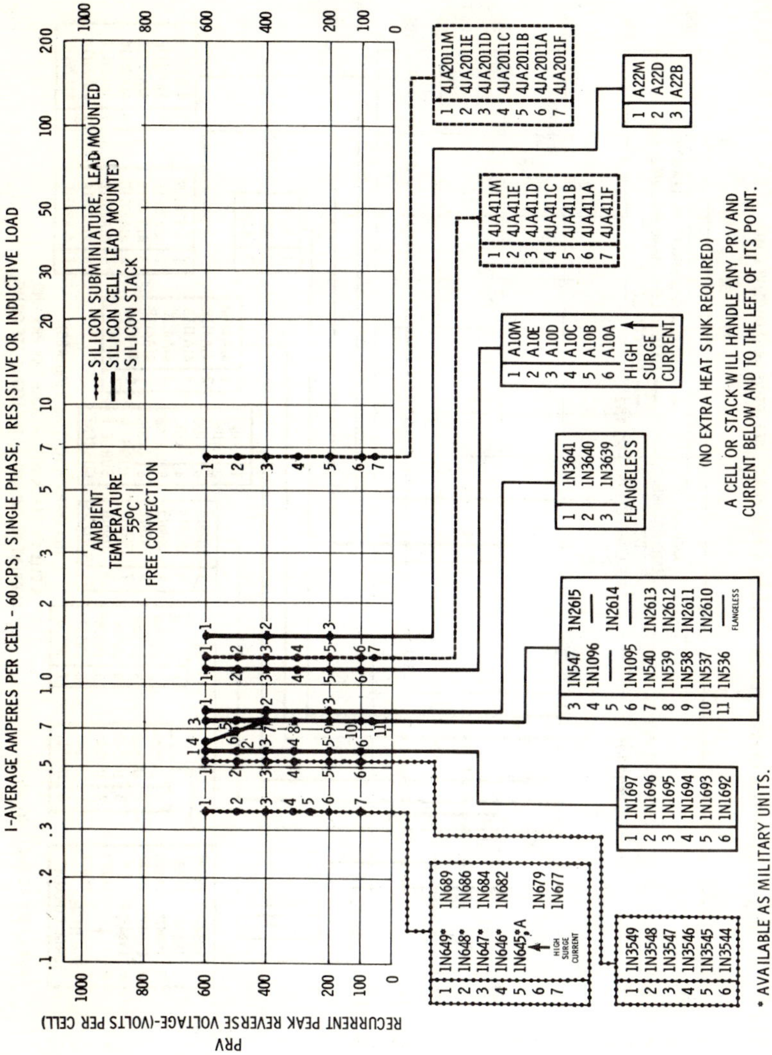

Courtesy General Electric Co.

Chart 3. Rectifier selection chart.

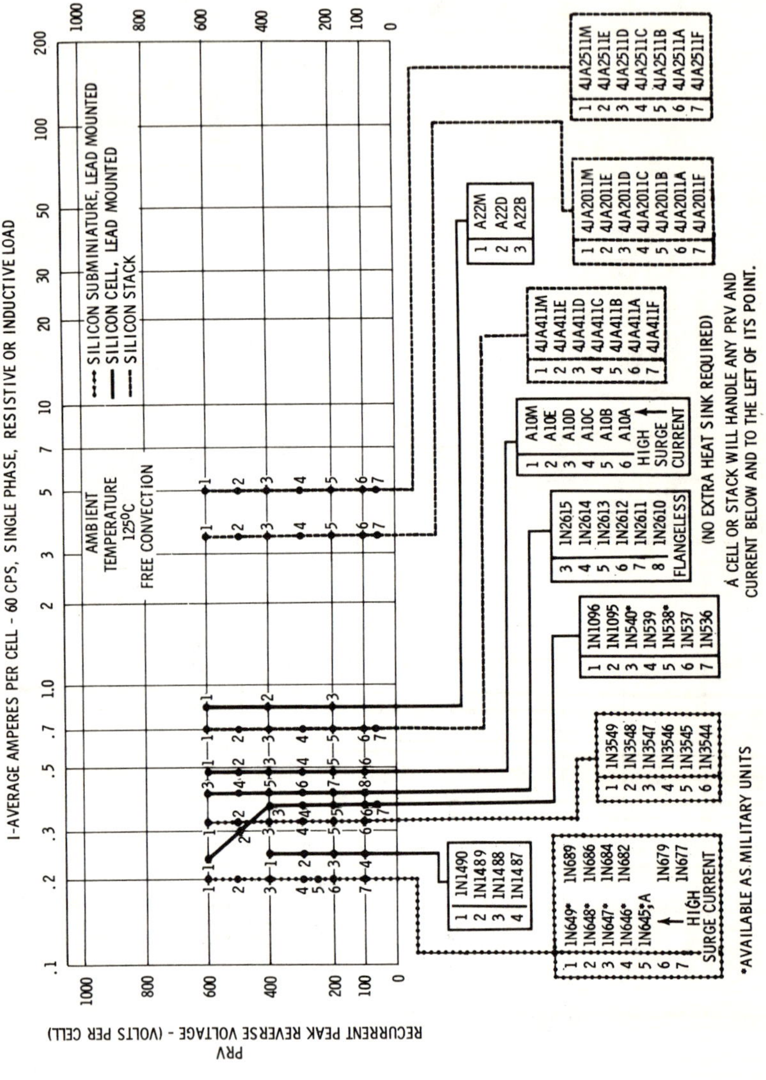

Courtesy General Electric Co.

Chart 4. Rectifier selection chart.

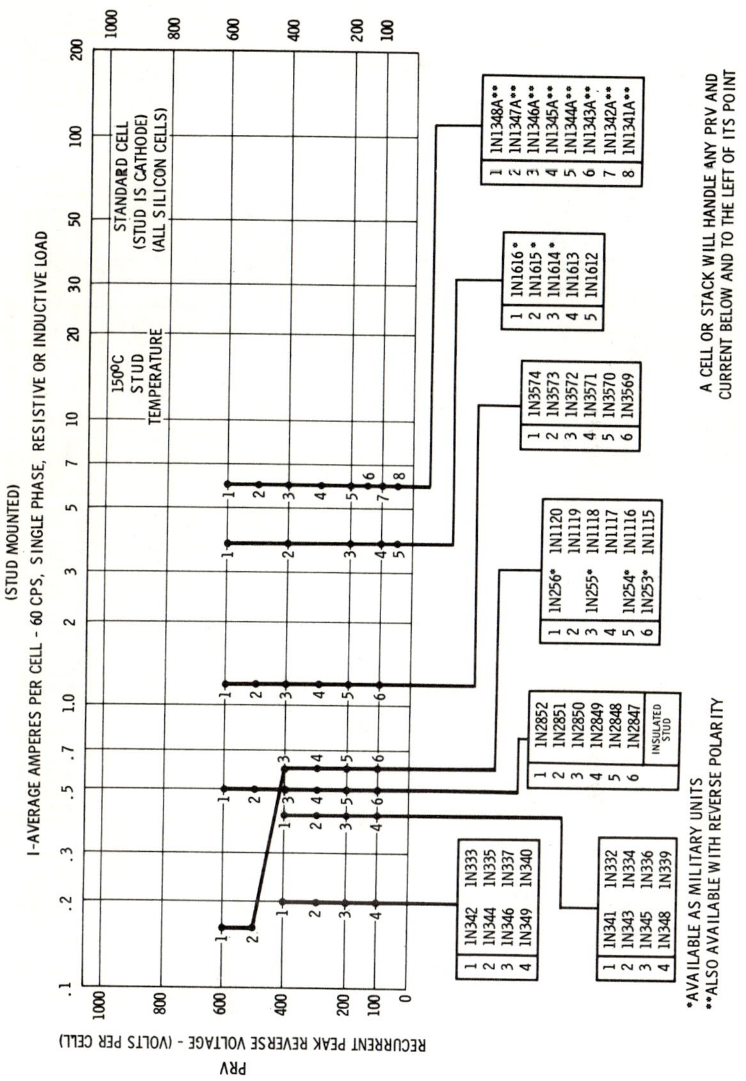

Courtesy General Electric Co.

Section 1

POWER SUPPLIES

UNREGULATED POWER SUPPLIES

Two different types of unregulated power supplies are shown here. In Fig. A, a full-wave bridge rectifier is shown using four silicon diodes, each diode having a 100-volt peak-inverse rating. There are four different ways that the circuit in Fig. A may be connected.

Using primary terminals 1 and 3 of the transformer with the secondary terminals connected in parallel and the choke terminals connected in parallel, the diode rating must be 1.5 amperes, and the resistance value is 70 ohms. Under these conditions, the output is 13 volts at 2.5 amperes with a 17% regulation.

With primary terminals 1 and 2 being used, and the secondary terminals of the transformer as well as the choke terminals connected in parallel, the diodes carry 1.5 amperes, and the resistance should now be 60 ohms. Under these conditions, the output is 16.5 volts at 2.5 ampere with a 12% regulation.

Using terminals 1 and 3 of the primary, with the secondary terminals connected in series and the choke terminals connected in series, each diode will conduct 0.75 ampere, and the resistance value

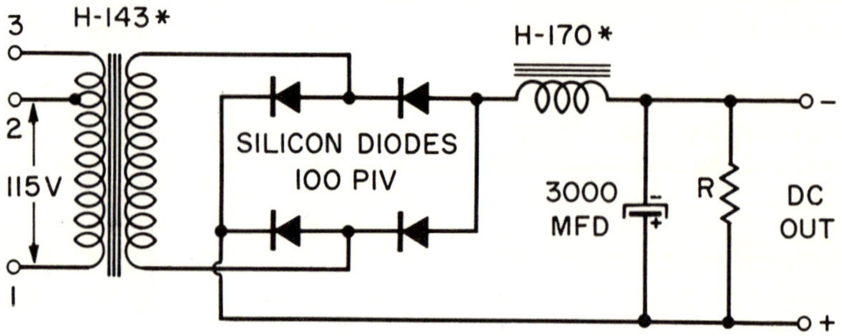

(A) Full-wave bridge rectifier.

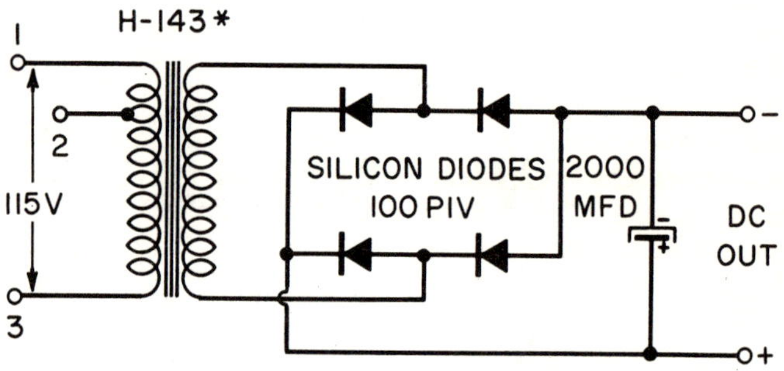

(B) Capacitive input filter.

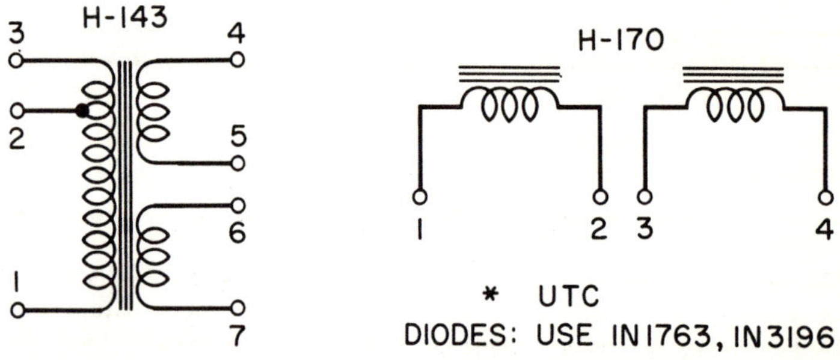

(C) Terminal connections of choke and transformer windings.

Courtesy United Transformer Corp.

Unregulated power supplies.

is now 300 ohms. The output is 27.5 volts at 1.25 amperes with a 13% regulation.

Using primary terminals 1 and 2, with the secondary terminals in series and the choke terminals in series, the diodes carry 0.75 ampere, and the resistance value is now 250 ohms. The output is 34.5 volts at 1.25 amperes with a 12% regulation.

Fig. B shows the same transformer and diodes, but with a capacitive input filter without a choke. Using primary terminals 1 and 3 with the secondary terminals connected in parallel, the diode will carry 1.5 amperes and the output is 19.6 volts at 2 amperes with a 30% regulation.

Using primary terminals 1 and 2, with the secondary terminals connected in parallel, the diode current is 1.5 amperes but the output is 25.5 volts at 2 amperes with a 26% regulation.

Using primary terminals 1 and 3 with the secondary terminals connected in series, each diode carries 0.75 ampere, and the output is 42 volts at 1 ampere with a 24% regulation.

Using primary terminals 1 and 2 with the secondary terminals connected in series, each diode carries 0.75 ampere, and the output voltage is 53 volts at 1 ampere with a 20% regulation.

Fig. C shows the terminal connections of the transformer windings and the choke windings.

BASIC LOW-VOLTAGE SUPPLIES

Three basic low-voltage power supplies are shown. Fig. A shows a half-wave supply, Fig. B shows a full-wave supply, and Fig. C shows a voltage doubler.

Fig. A is a simple half-wave circuit. With a power transformer with a 2-to-1 step-up operating from a 117-volt line, this circuit produces from 270 to 290 volts dc.

The full-wave circuit shown in Fig. B is essentially two of the circuits shown in Fig. A; the circuit can deliver an output voltage of 280 to 310 volts, depending on the current drain required of the circuit.

For both circuits the suggested rectifiers are the F-8, which has a maximum peak-inverse voltage of 800, a maximum rms voltage of 560, and a maximum d-c current of 750 ma. The ST18 or the 80H may also be used. All of these rectifiers have the same ratings, but they are different physical types.

Fig. C shows a voltage doubler operating with a 1-to-1 transformer and producing from 270 to 290 volts dc. Suggested rectifiers are the

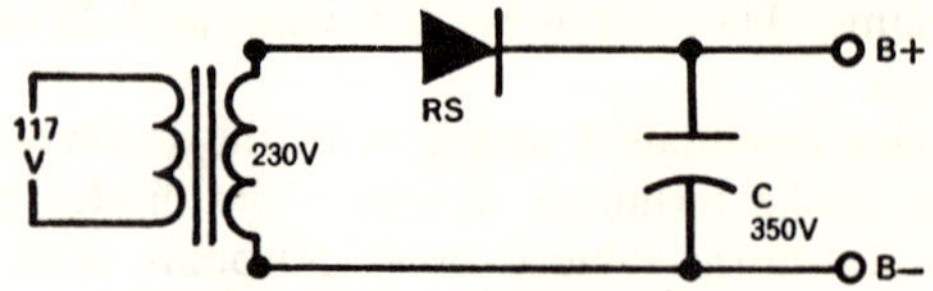

RS=Sarkes Tarzian F-8, ST18, or 80H

Output:

Volts DC	MA DC	C
290	250	150 Mfd
280	500	200 Mfd
270	750	200 Mfd

(A) Half-wave power supply.

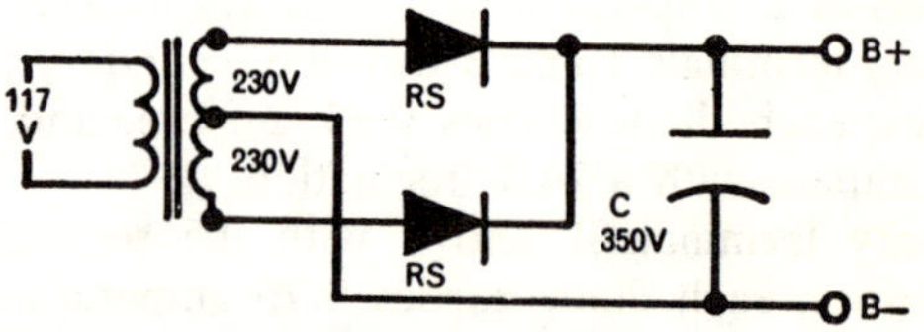

Each RS=Sarkes Tarzian F-8, ST18 or 80H

Output:

Volts DC	MA DC	C
310	500	50 Mfd
290	1,000	50 Mfd
280	1,500	50 Mfd

(B) Full-wave power supply.

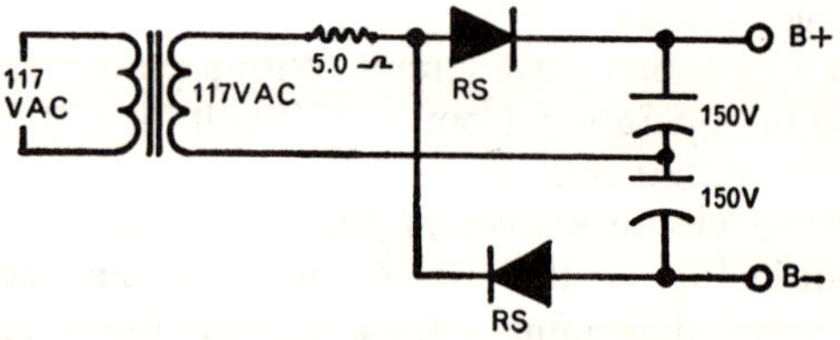

Each RS=Sarkes Tarzian 40H, F-4 or ST14

Output:

Volts DC	MA DC	Each C
290	300	100 Mfd
280	500	100 Mfd
270	750	100 Mfd

(C) Full-wave voltage doubler.

Courtesy Sarkes Tarzian Inc.

Basic low-voltage supplies.

40H, the F-4, or the ST14. These rectifiers have a maximum peak-inverse voltage of 400 volts, a maximum rms voltage of 280, and a maximum d-c current of 750 ma.

REGULATED POWER SUPPLY

This circuit shows a regulated power supply, with a 115-volt input to terminals 1 and 3 of the transformer primary. The secondary windings are connected in parallel and feed a bridge rectifier with a capacitive filter.

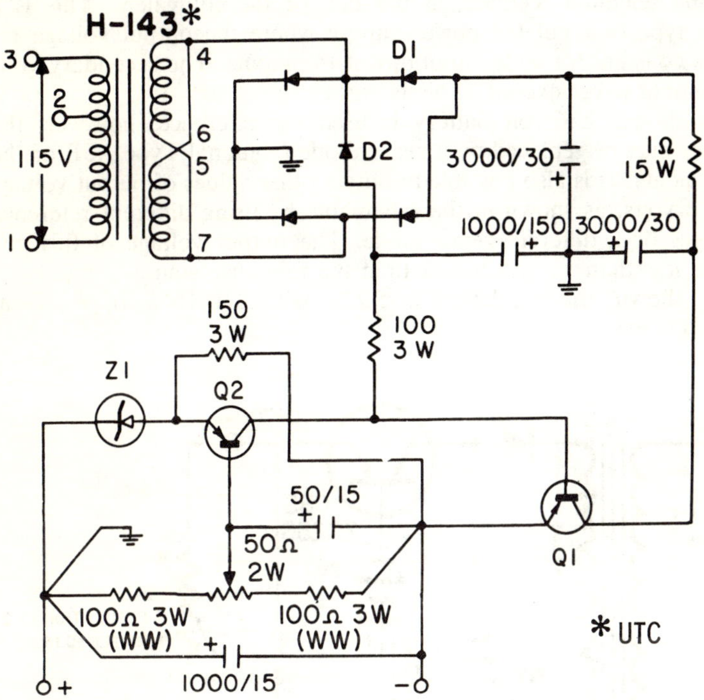

D1 (4) 1.5A 100 PIV IN1613 D2 (1) .75A 100 PIV IN440B
Z1 5.6V ZENER (IN1520)
Q1 DELCO 2N441 (SINK 7270725)
Q2 RCA 2N301 (HEAT SINK)

Courtesy United Transformer Corp.

Regulated power supply.

A regulator is used to correct the line-voltage variations or load variations. The output voltage can be varied from 11.5 to 12.5 volts dc by means of the control in the base of transistor Q2. Output current can vary from 0 to 2 amperes; there is 5-millivolt ripple in the output. The regulation for a line voltage from 105 to 125 volts is 3.5%.

REGULATED SUPPLY

This is a simple regulated power supply operating from 115 volts ac and using a single power transistor. The rectifier diode (D_1) is two 500-ma selenium rectifiers in parallel, or the equivalent. This is a series type of regulated power supply where the output voltage remains constant for wide variations of the inputs. The secondary voltage should never exceed 25 volts rms.

As shown, a 6-volt battery is used for reference; however, the battery may be replaced by a zener diode (National Type A5B or the equivalent). It is also possible to obtain other values of output voltage than 6 volts, as shown in the schematic, by using different reference voltages or a different zener diode. The output voltage of this circuit is less than ½ volt higher than the reference source.

As shown, the regulation is 5.2% with a 0.16% ripple, from 0 to 700 ma.

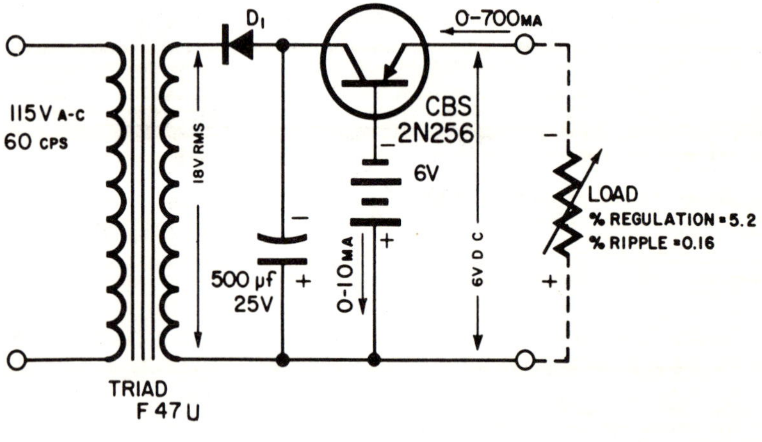

Courtesy CBS Electronics Div., CBS Inc.

Regulated supply.

REGULATED POWER SUPPLY

This circuit shows a regulated power supply, with a 115-volt input to terminals 1 and 2 of the transformer. Output is variable from 26 to 30 volts dc; for the adjustment use the control in the base of transistor Q3.

After it is filtered, the output from the bridge rectifier is fed to the regulation circuit which compares the output voltage to a reference voltage. The positive end of the output is grounded, and the negative output voltage can be varied from 26 to 30 volts. The ripple in the output is 0.12 millivolt. Regulation with a line voltage variation from 105 to 125 volts is 0.1%.

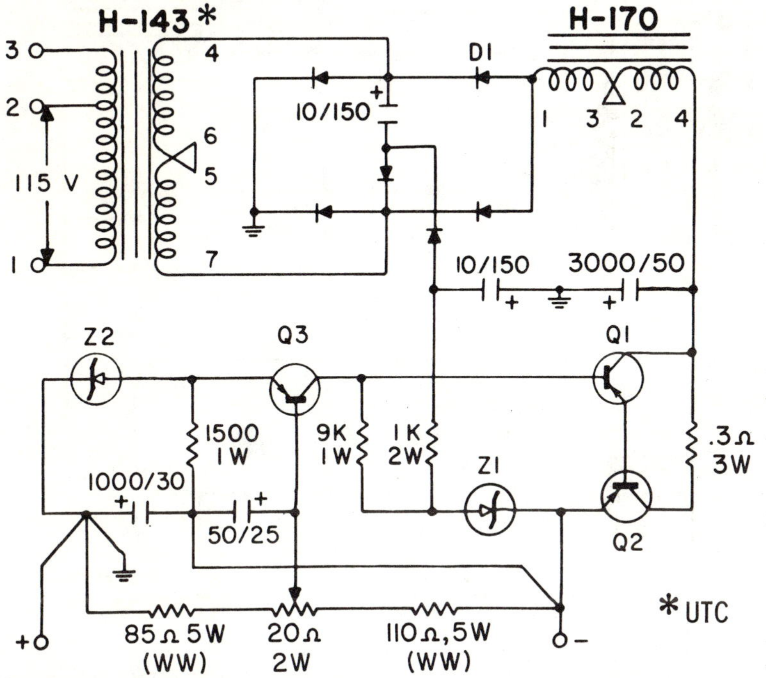

D1 (6) .75A 100 P.I.V. (IN440B)
Z1 24V ZENER (IM24Z) , Z2 12V ZENER (IN1513)
Q1 PHILCO 2N671 (HEAT SINK)
Q2 DELCO 2N441 (SINK 7270725)
Q3 TEXAS 2N1377

Courtesy United Transformer Corp.

Regulated power supply.

REGULATED POWER SUPPLY

This circuit shows a regulated power supply fed from a 115-volt line; the output is from 11 to 13 volts dc (the positive end is grounded) with a current of 0 to 2 amperes.

A bridge rectifier is used with a choke and a filter capacitor. The regulator circuit compares the output voltage with the reference voltage and compensates for any variations of load or line. There is a ripple of 0.35 millivolt. With a line voltage variation of 105 to 125 volts, the regulation is 0.1%. Note the special nature of the ground on the positive side; it is 4 inches of No. 22 solid wire.

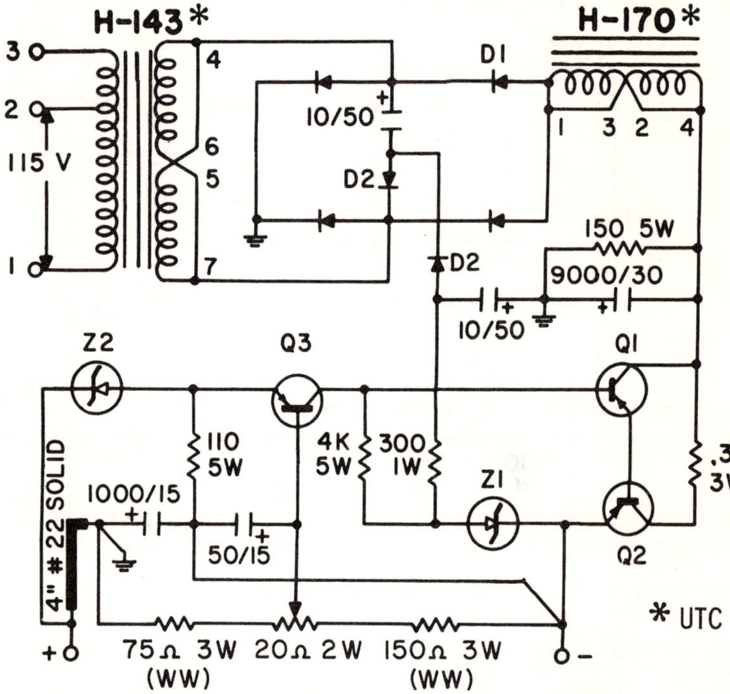

D1 (4) 1.5A, 100 P.I.V. - 1N1613 D2 .75A, 100 P.I.V.- 1N440B
Z1 12V ZENER (1N1513) Z2 3.9V ZENER (1N1507)
Q1 PHILCO 2N671 (HEAT SINK)
Q2 DELCO 2N441 (SINK 7270725)
Q3 TEXAS 2N1377

Courtesy United Transformer Corp.

Regulated power supply.

REGULATED POWER SUPPLY

This circuit shows a regulated power supply with an output of 27 to 29 volts dc at up to 1 ampere. Line-voltage input is between point 1 and point 3 of the transformer; the secondary is connected in series. A capacitive filter is used. The regulating circuit compares a reference voltage to the output and provides for proper compensation to keep the voltage output constant.

The ripple in the output is 1.5 millivolts; there is a 1% regulation for variations in the line input from 105 to 125 volts. An adjustment is provided in the base of transistor Q2 to adjust for the desired power output.

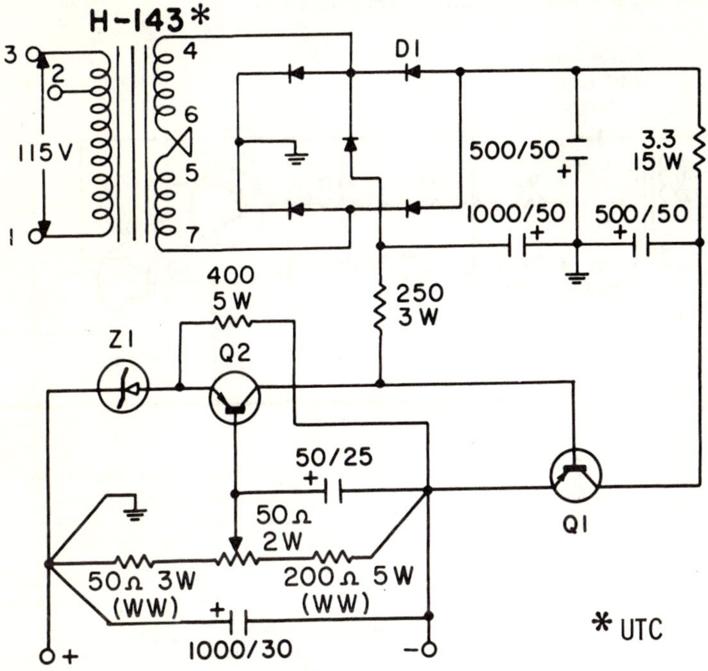

D1 (5) .75A 100 PIV (IN440B)
Z1 6.8V ZENER (1.5M 6.8Z)
Q1 DELCO 2N441 (SINK 7270725)
Q2 RCA 2N301 (HEAT SINK)

Courtesy United Transformer Corp.

Regulated power supply.

12-VOLT REGULATED SUPPLY

This circuit shows a 12-volt regulated power supply using transistor types 2N1274 and TI3027. The 2N1274 is a germanium small-signal transistor, while the TI3027 is a power transistor.

In this circuit the 1N759 is a reference diode connected to the base of the 2N1274. The 2N1274, connected by a Darlington connection to the TI3027, acts as a series regulator. This regulator is extremely sensitive to current changes in the reference diode.

The change in output voltage from no-load to a full 1.5-ampere load is 0.4 volt, while the ripple at full load is 30 millivolts. The change in the output voltage with a 10% change in the line voltage from the normal 117 volts is very small.

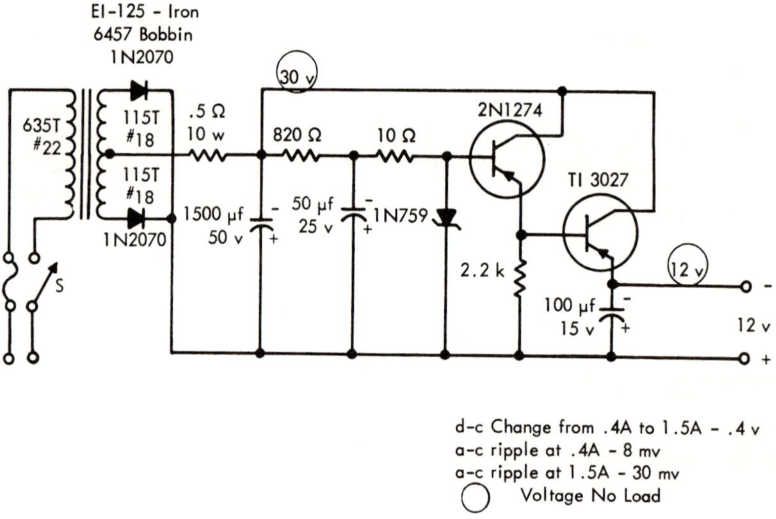

Courtesy Texas Instruments Inc.

12-volt regulated supply.

450/800-VOLT SUPPLY

This is a dual-voltage power supply that provides two voltage outputs; one output is 800 volts at 175 ma with a 1% ripple at a 16% load regulation, such as is required for a final radio-frequency power amplifier, and a second output is 450 volts at 25 ma with a 0.02% ripple, such as is required for a preamplifier or an oscillator circuit.

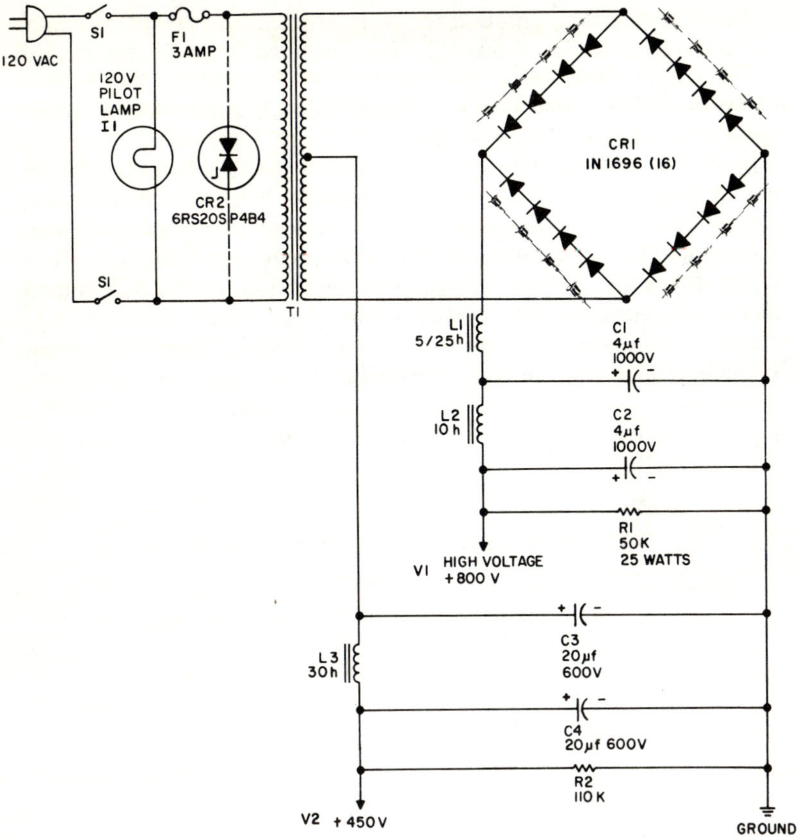

C_1, C_2—4 mfd, 1000 volt capacitor (Cornell-Dubilier 10040)

C_3, C_4—20 mfd, 600 volt electrolytic capacitor

CR_1—16 G-E type 1N1696 silicon rectifier diodes connected in groups of four

CR_2—G-E type 6RS20SP4B4 Thyrector diode (optional transient voltage suppressor)

F_1—3 AGC fuse, 3 amps

I_1—120 volt, 6 watt pilot lamp

L_1—5/25 henry choke, 175 ma (UTC S-30, or equivalent)

L_2—10 henry choke, 175 ma (UTC S-29, or equivalent)

L_3—30 henry choke, 25 ma (UTC S-25, or equivalent)

R_1—50K, 25 watt resistor

R_2—110K, 4 watt resistor (2-220K, 2 watt resistors in parallel)

S_1—DPST switch

T_1—200 ma transformer: primary, 120 volt AC, 60 cps; secondary, 800 volt (Stancor PC-8412, or equivalent)

Courtesy General Electric Co.

450/800-volt supply.

There is a common ground for both outputs. To provide a variable d-c voltage output, an adjustable line transformer can be used before the step-up transformer.

As shown, the transformer is center-tapped, and a bridge rectifier with four silicon diodes in each leg is used for the high-voltage output. The same transformer is used with the rectifier diodes in the two right-hand legs to provide a full-wave rectifier for the lower-voltage power supply. The high-voltage power supply uses a choke-input filter, while the low-voltage power supply uses a capacitor-input filter. If desired, the 450-volt supply may be reduced to 375 volts by removing capacitor C3 from the circuit.

A transient-suppressor diode (CR-2) is used to protect the silicon rectifiers against the voltage transients; a pilot lamp is also shown, and it is suggested as a safety precaution, since in the absence of filaments it is not always possible to tell from inspection when the power supply is on.

TRANSFORMERLESS D-C POWER SUPPLY

This circuit shows a transformerless d-c power supply suitable for use in lightweight equipment; a single controlled rectifier is used with-

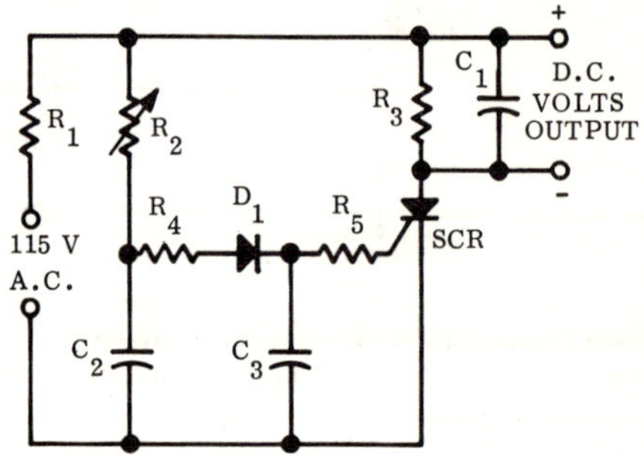

Parts List

R_1	- 1 Ω (1W)	C_1	- 1000 μf(15V)
R_2	- 10 KΩ	C_2	- 1.5 μf
R_3	- 5 KΩ	C_3	- 15 μf
R_4	- 1 KΩ	D_1	- 1N4001
R_5	- 470 Ω	SCR	- MCR1304-4

Courtesy Motorola Semiconductor Products

Transformerless d-c power supply.

out a unijunction transistor. Alternating current from the line source is applied to the circuit as shown. There is a phase delay because of the two circuits, one of which is resistor R_4 and capacitor C_2, while the other is resistor R_5 and capacitor C_3. The rectifier diode is used to prevent reverse current flow to the gate.

As resistor R_2 is varied, the voltage which is phase-shifted to the gate of the controlled rectifier is changed; hence, the on time of the controlled rectifier can be varied. Capacitor C_1 is the output filter capacitor. At 100 ma of output current, this circuit provides 10 volts output; at 200 ma of output current the output voltage drops to approximately 6 volts.

REGULATED A-C POWER SUPPLY

This circuit shows a regulator for a-c power, and it performs the same function as a constant-voltage transformer at considerably less cost and in much less space. The line voltage is in series with the load and the regulator, as shown. Resistor R_1 is used for adjustment to allow the peak voltage point to increase when the line voltage is increased, thereby retarding the firing angle of the controlled rectifiers and causing a reduction in the voltage applied to the load.

Resistor R_2 is used to adjust the circuit for any desired output voltage. Current output is determined by the rating of the controlled rectifiers, and for the rectifiers shown in the circuit, the current can

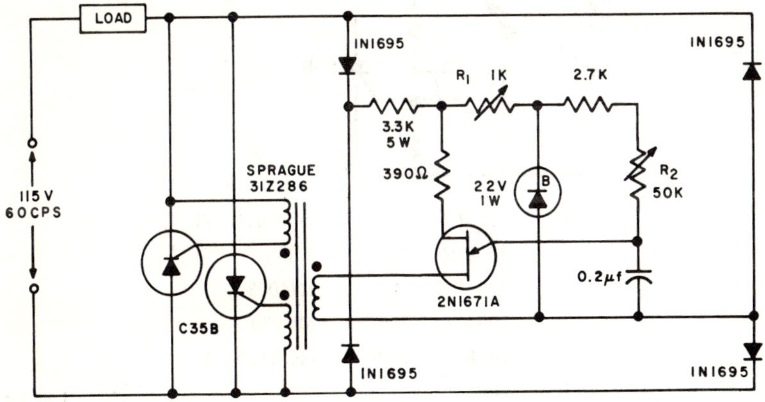

Courtesy General Electric Co.

Regulated a-c power supply.

be as much as 32 amperes. Functionally, a bridge rectifier is used to provide d-c power for unijunction-transistor 2N1671A, whose output is fed to the gate leads of the controlled rectifiers.

TRANSFORMERLESS D-C POWER SUPPLY

This circuit shows a lightweight power supply designed to produce a 10-volt output at 100 ma from a standard line voltage source. A controlled rectifier and unijunction transistor are used. As shown in the circuit, resistors R_2 and R_3, as well as capacitor C_2, are used to make up the time delay, which is a part of the unijunction transistor circuit; this time delay determines the unijunction switching. If the value of resistor R_3 is increased, the output voltage of the circuit decreases. The trigger pulses from the unijunction transistor are applied to the gate of the controlled rectifier. The rectifier is then turned off and on by the unijunction transistor, and the chopped ac is filtered by means of capacitor C_1. When this capacitor is 1000 microfarads, as shown in the circuit, a 2.5% ripple voltage will appear across the load resistance at the designed center value of 10 volts and 100 milliamperes. Total power-supply efficiency is about 77%.

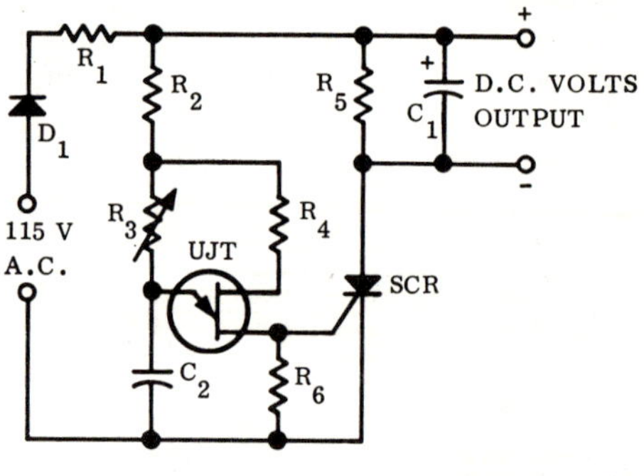

R_1	-	1 Ω (1 W)	C_1	-	1000 μf(15V)
R_2	-	39 KΩ	C_2	-	0.5 μf
R_3	-	100 KΩ	D_1	-	1N4003
R_4	-	390 Ω	UJT	-	MU970
R_5	-	10 KΩ	SCR	-	MCR1304-4
R_6	-	100 Ω			

Courtesy Motorola Semiconductor Products

Transformerless d-c power supply.

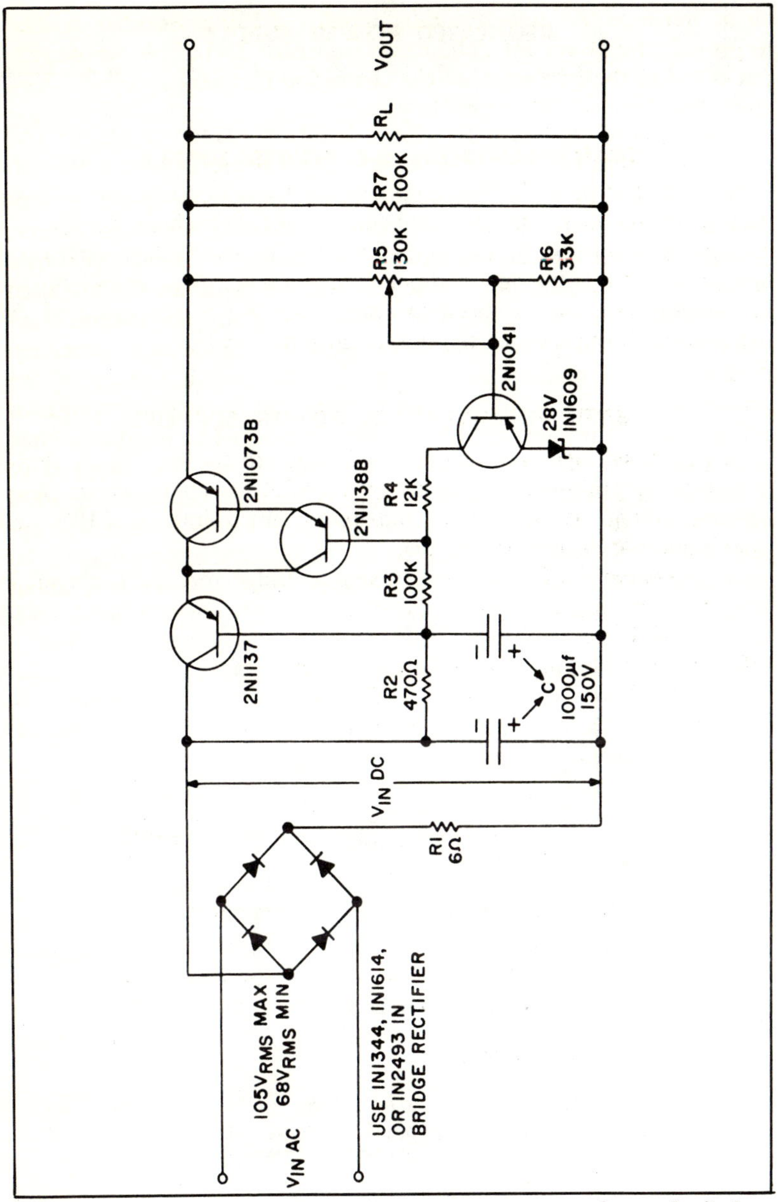

Courtesy Bendix Corp.

Regulated power supply.

REGULATED POWER SUPPLY

This circuit shows a regulated power supply using rectifier types 1N1344, 1N1614, or 1N2493.

A-c input to the bridge rectifier is 105 volts maximum and 68 volts minimum; both of these voltages are rms values. Output is 90 volts at a current of up to 700 milliamperes. A transistor regulator follows the bridge rectifier. This is a conventional rectifier where the 2N1041 compares a portion of the output voltage to the voltage developed across the 1N1609 zener diode; this voltage, by means of the 2N1138B transistor, is used to control the series resistance of the transistors through which the output current flows.

BASIC HIGH-VOLTAGE POWER SUPPLIES

Three basic high-voltage power supplies are shown. Fig. A shows a half-wave 1000-volt supply using the S5521 rectifier whose peak-inverse voltage is 3000 volts, maximum rms voltage is 2100, and maximum d-c current is 150 ma.

A full-wave 5000-volt supply using a bridge rectifier is shown in Fig. B. The suggested silicon diode is the S-5531 which has a maximum peak-inverse voltage of 10,000 volts, a maximum rms value of 7000 volts, and a maximum d-c current of 25 ma.

Fig. C shows a three-phase 7500-volt supply with a full-wave rectifier in each leg. The suggested rectifier is the same as the one used in the circuit in Fig. B.

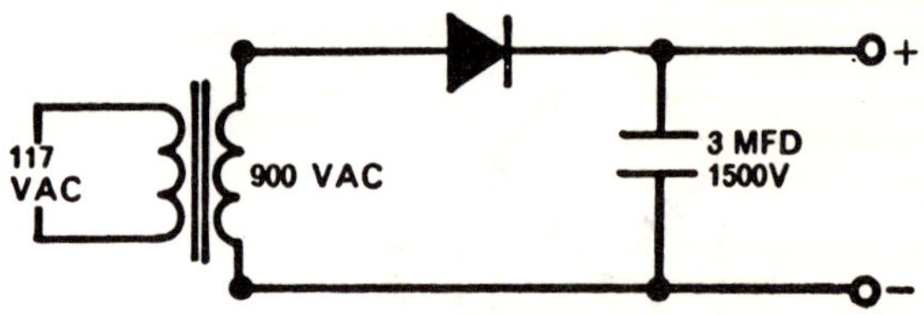

RS=Sarkes Tarzian S-5521

Output:

Volts DC	MA
1,000	120 ma @ 25 degrees C
1,000	60 ma @ 100 degrees C

(A) Half-wave power supply.

Basic high-voltage

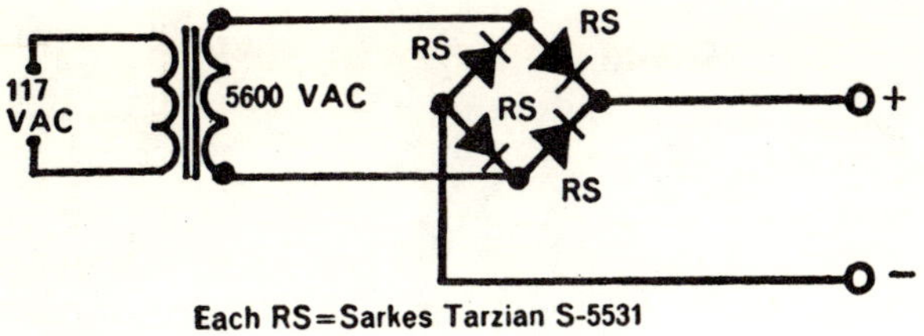

Output:
5000 VDC @ 150 ma
@ 25 degrees C

5000 VDC @ 50 Ma
@ 100 degrees C

(B) Full-wave power supply.

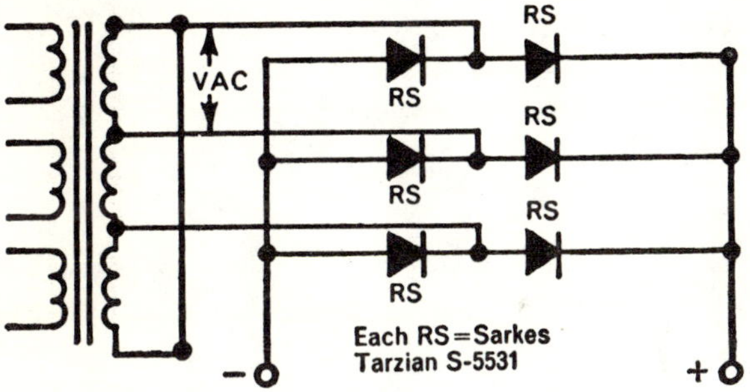

Input:
5700 Volts AC per
phase

Output:
7500 VDC @ 200 ma
@ 25 degrees C
7500 VDC @ 65 ma
@ 100 degrees C

(C) 3-phase power supply.

Courtesy Sarkes Tarzian Inc.

power supplies.

Section 2

VOLTAGE REGULATORS

VOLTAGE REGULATOR WITH OVERLOAD PROTECTION

This circuit shows a voltage regulator with overload protection for the series transistor. A short circuit in the load will cause the series transistor to act as the load in most voltage-regulator circuits, and this can damage the transistor under these conditions. Because the series transistor is directly in the current path of the load, protection is required in a very short time. Three methods of protection are: (1) protection by load switching, (2) protection by current interruption, and (3) protection by current limiting.

This circuit shows current interruption, which is a dependable method of protection where the overload can be detected and interruption of the load circuit can be obtained before there is damage to the transistors acting as regulators. The circuit shows a complete regulator with a 115-volt input to a 2-to-1 stepdown power transformer; output of this regulator is 60 volts.

In this circuit the 2N1601 controlled rectifier will conduct within one microsecond after a short circuit is placed across the regulator load. The 1-ampere fuse will open within three to four milliseconds.

In the same type of circuit without the controlled-rectifier protection, the same 1-ampere fuse would need about 400 milliseconds in order to open, and by this time the transistor would probably be damaged. Thus, this circuit provides regulation with protection for the series transistor, which is the 2N389. The series resistance that this transistor presents to the circuit depends on the 2N656 which amplifies the error voltage obtained from the 2N343B. The 2N343B compares a portion of the output voltage with the voltage developed across the 1N754 zener diode.

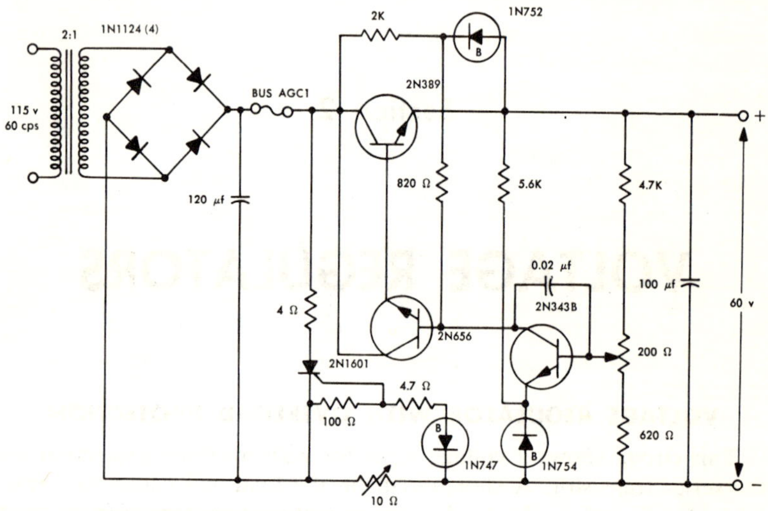

Courtesy Texas Instruments Inc.

Voltage regulator with overload protection.

SWITCHING VOLTAGE REGULATOR

A proportional voltage regulator is shown in Fig. A. In this circuit a voltage proportional to the output is compared to a voltage of the reference diode (CR_4) in the emitter circuit of transistor Q_2. A voltage proportional to the output voltage is obtained across the tap on resistor R_3. A differential signal is amplified by transistor Q_3, and the output signal of this differential amplifier is used to control the current in series transistor Q_1. Q_1 is effectively a variable series resistance. Thus, for this circuit a 17-volt input will provide a regulated output of 12 volts at 2 amperes.

Fig. B effectively shows the same circuit except that an inductance (L_1), a capacitance (C_2), and a diode (CR_2) are added to the circuit; these additions change the circuit into a switching type of regulator. If in Fig. B capacitor C_2 is disconnected, there is no feedback, and the remaining circuit will be a proportional voltage regulator. When, however, this capacitor is connected in the circuit, the operation is in the switching mode since there is a positive feedback loop and oscillations will start. The base current of transistor Q_2 has two individual components—one is the charge and discharge of capacitor C_2, while the second component is proportional to the difference between voltages V_1 and V_2. Transistors Q_2 and Q_3, together with capacitor C_2, form a free-running multivibrator. This multivibrator, in effect, turns on and off transistor Q_1, which is the series transistor; in effect, this is the switching mode of operation for this regulator.

Filter reactance L_1 is used to isolate transistor Q_1 from filter capacitor C_1 so that voltage-saturated operation is possible.

Resistor R_1 determines the base current of transistor Q_1, and the value for this resistor is determined by dividing the input voltage by the base current for maximum load. Resistor R_2 is part of the time-constant network that controls the frequency of oscillation, while resistor R_3 is used for adjustment of the output voltage. Resistor R_4 provides a bias current to the voltage-reference diode CR_4, while resistor R_5 limits the current through transistor Q_2 if there is a short in transistor Q_1.

Capacitor C_1 is a filter capacitor, capacitor C_2 is a feedback capacitor which determines the frequency of oscillation, while capacitor C_3 controls the frequency of oscillation as well as having the function of suppressing any parasitic oscillations.

There are four diodes in this circuit; CR_1 provides bias for transistor Q_3 and allows Q_3 to apply reverse bias to Q_1; CR_2 is a clamping diode while CR_3 reduces the leakage in transistor Q_3. CR_4 is the voltage-reference diode and this diode has a positive temperature coefficient.

The regulation characteristics for this circuit are shown in Fig. C.

SERIES VOLTAGE REGULATOR

This circuit shows three transistors in a series voltage-regulator circuit; two of these transistors are type 2N1481, while one is a type 2N1489. In series voltage regulators of this type, a cascade d-c amplifier is used to amplify a difference signal, or an error, which is obtained by means of a comparison between part of the output volt-

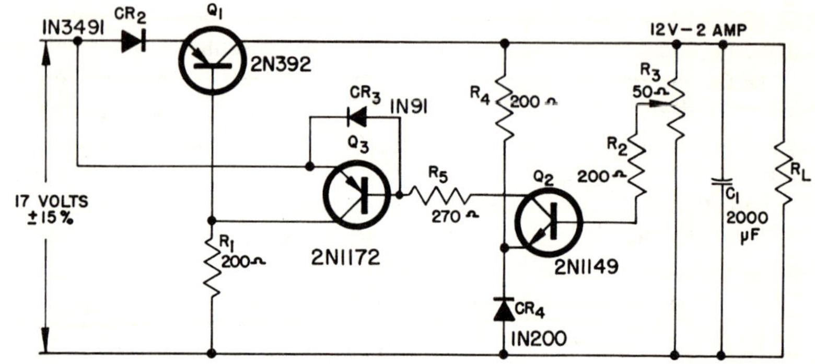

(A) Proportional voltage regulator.

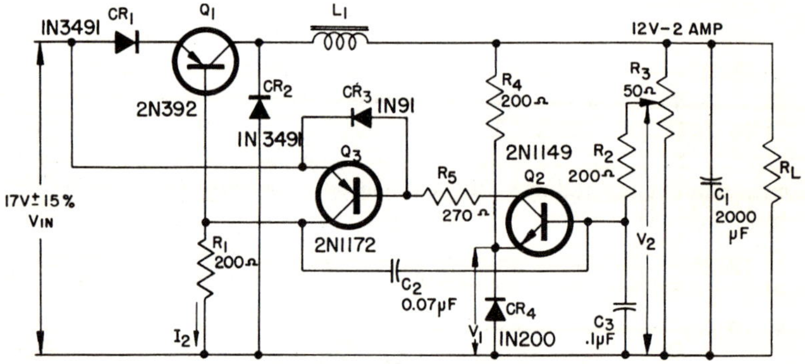

L_1 CORE EI 21-1/2", STACK .030" AIR GAP, 120T No. 20 WIRE

(B) Switching voltage regulator.

Voltage

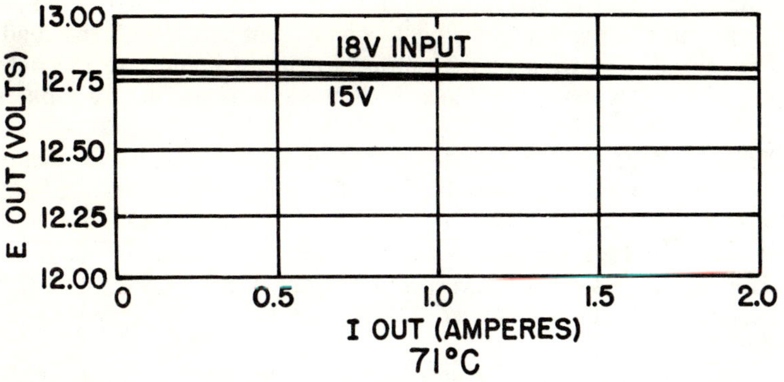

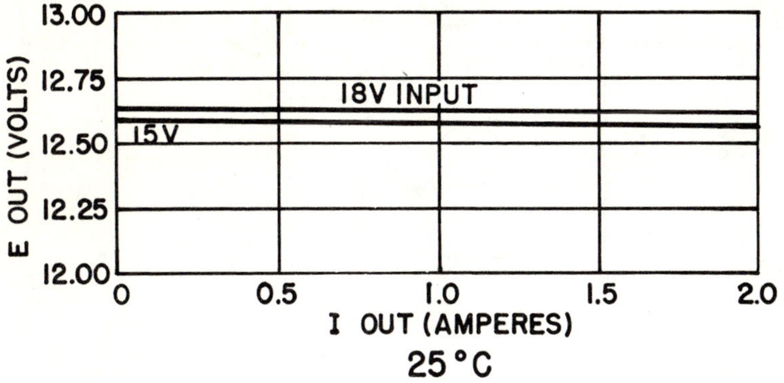

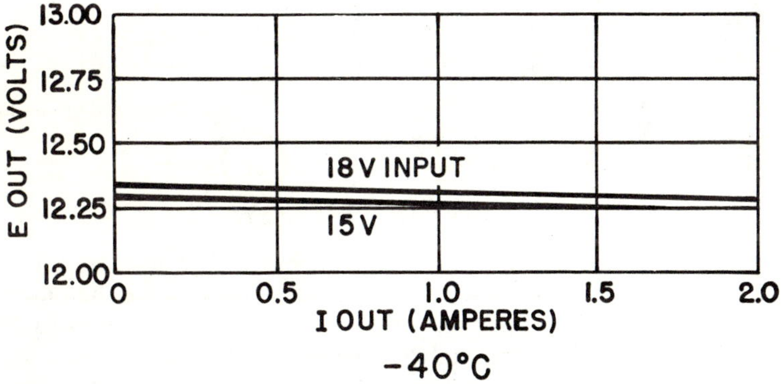

(C) Regulation characteristics.

Courtesy Delco Radio Div., General Motors Corp.

regulators.

age and a reference voltage. The amplified error signal is used as the input to the regulating element, which is in series with the load. In this circuit a portion of the output voltage is applied to the base of transistor Q1, while the zener 12-volt diode is tied to the emitter of the same transistor.

Transistor Q2, then, acts as a differential, or error, amplifier whose input signal is the difference between the reference voltage and the voltage which is proportional to the output. Transistor Q2, in effect, controls series transistor Q3, which provides the necessary regulation.

This circuit, as shown, accepts a voltage input of 45 volts, ±5 volts, and produces a regulated 28-volt output.

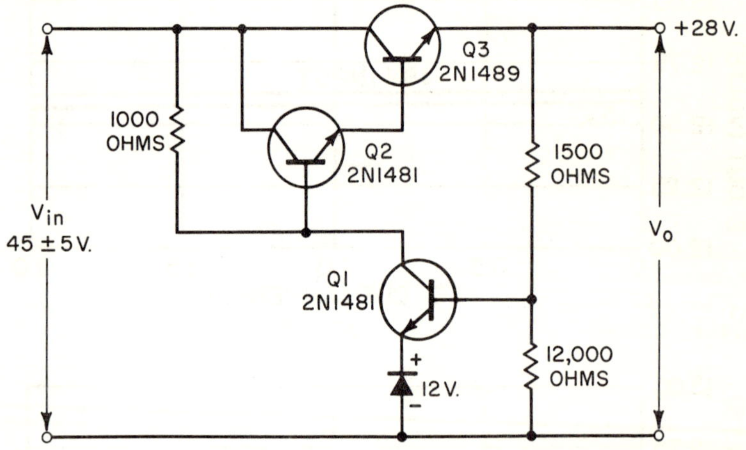

Courtesy Radio Corporation of America

Series voltage regulator.

SERIES VOLTAGE REGULATOR

This circuit shows a series type of voltage regulator with an adjustable output. Input to this circuit is 40 to 50 volts dc, while the output can be adjusted from 22 to 30 volts dc for a current from 0 to 10 amperes. Line regulation is within 1%, and load regulation is within 0.5%. A 12-volt reference diode is used to provide a voltage which is compared to a portion of the output voltage; this difference, or error, voltage is amplified and used to control the three series transistors type 2N2016.

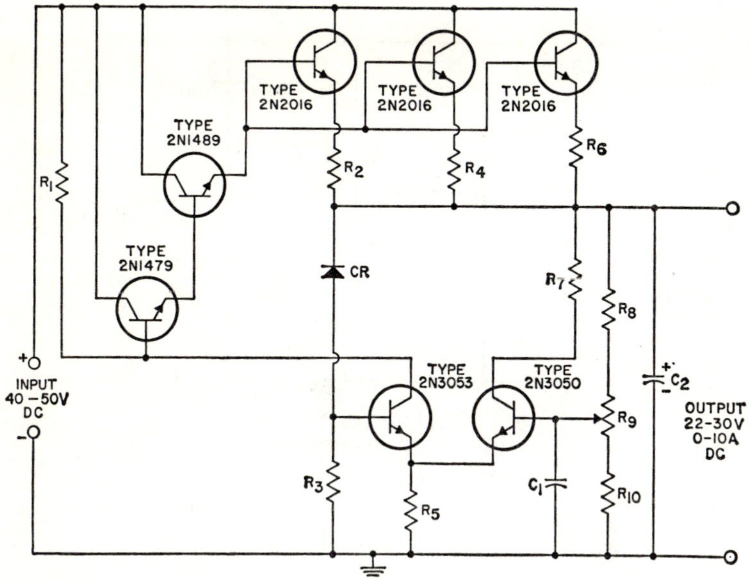

C_1—1 mfd, 25V, paper
C_2—100 mfd, 50 V electrolytic
CR—Reference diode, 12 V
R_1—1200 ohms, ½ watt
R_2, R_4, R_6—0.1 ohm, ½ watt
R_3—2K, ½ watt
R_5—570 ohms, ½ watt
R_7—270 ohms, ½ watt
R_8, R_{10}—1K, ½ watt
R_9—1K, ½ watt potentiometer

Courtesy Radio Corporation of America

Series voltage regulator.

SHUNT VOLTAGE REGULATOR

This circuit shows a two-transistor shunt voltage regulator using one transistor, type 2N1485, and one type 2N1481. Shunt regulators of this type are less complex than series regulators, although they are not quite as efficient. A shunt regulator uses a voltage-reference element and a shunt element. In this circuit the output voltage remains constant because the shunt-element current changes when either the input voltage or the load current changes. The change in the shunt current appears as a change in voltage across the resistor that is in series with the load.

This circuit has an input of 49 volts ±7 volts, and an output of 28 volts as shown. The series resistance required in this case is 18 ohms.

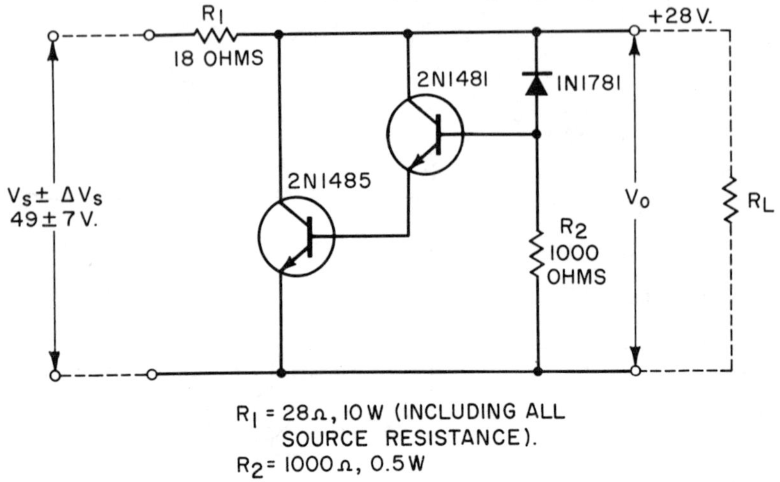

Courtesy Radio Corporation of America

Shunt voltage regulator.

SERIES VOLTAGE REGULATOR

This circuit shows a simple three-transistor series voltage regulator with a voltage regulation of 3.5%. The input to this circuit is from 40 to 50 volts dc, while the output is 28 volts dc, with a current up to 1 ampere. A 12-volt reference diode is used to provide a voltage which is compared to a portion of the output voltage. The difference voltage is then amplified by the 2N1481 and applied to the base of the series transistor 2N1489.

CR = reference diode, 12 v.
R_1 = 1000 ohms, 1 watt
R_2 = 15000 ohms, 0.5 watt
R_3 = 12000 ohms, 0.5 watt

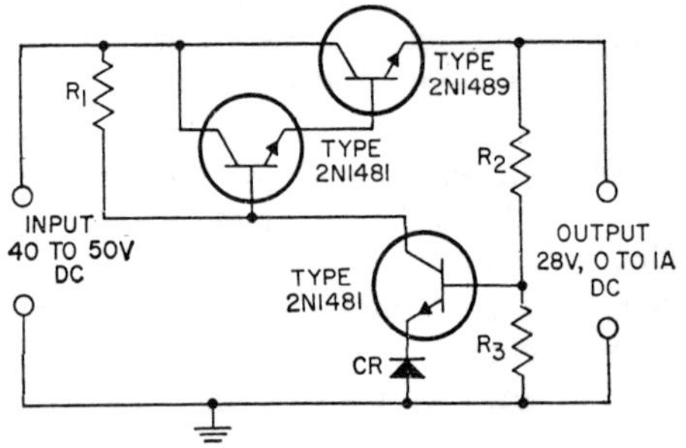

Courtesy Radio Corporation of America

Series voltage regulator.

SERIES VOLTAGE REGULATOR

This circuit shows six transistors used in a series type of voltage regulator to produce regulation within 2.5%.

The input is from 75 to 160 volts dc, and the output is 70 volts for a current from 0 to 4 amperes. A 33-volt reference diode is used to provide a comparison to produce the difference voltage; this voltage is amplified and used to control the three transistors type 2N1489 through which the load current is drawn.

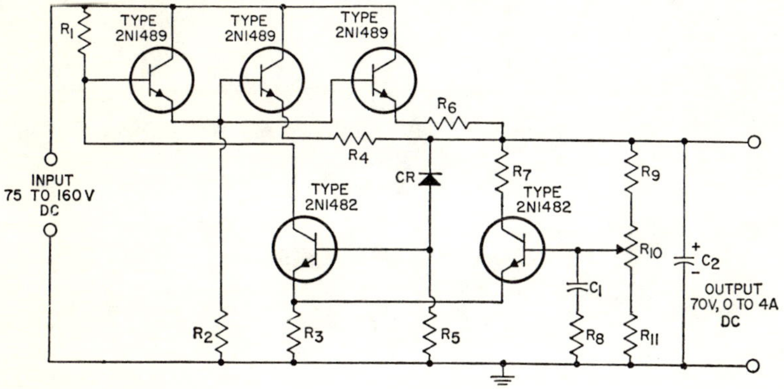

C_1—0.5 mfd, 50 V, paper
C_2—25 mfd, 100 V, electrolytic
CR—Reference diode, 33V
R_1—620 ohms, 1 watt
R_2—20K, ½ watt
R_3—1K, 1 watt
R_4; R_6—0.5 ohm, 10 watts
R_5—4K, ½ watt
R_7—750 ohms, ½ watt
R_8—100 ohms, ½ watt
R_9—2500 ohms, ½ watt
R_{10}—1K, ½ watt potentiometer
R_{11}—3500 ohms, ½ watt

Courtesy Radio Corporation of America

Series voltage regulator.

Section 3

CONVERTERS

SERIES-CONNECTED DC-TO-DC CONVERTER

This circuit shows a series connected dc-to-dc converter using four power transistors type 2N1100. In addition to the schematic diagram, there are three curves showing how voltage, frequency, and efficiency vary for three different ambient temperatures.

A series connection of transistors allows the operation of converter circuits such as this at higher supply voltages than would ordinarily be possible. All of the transformer windings must be on the same core in order to provide for proper phasing and equal power distribution so that the voltages applied to each transistor will be equal. These transistors act as switches controlled by bias voltages; the bias voltage for each transistor is obtained from the feedback winding. With no voltage induced in the feedback winding, resistors R_1, R_2, R_3, and R_4 (as shown in the circuit) act as voltage dividers to provide enough forward bias to allow the beginning of conduction and oscillation. For explanation, assume that transistors Q_1 and Q_3 are conducting, while transistors Q_2 and Q_4 are not conducting; the supply voltage is thus divided to the left-hand half of each of the primary windings as shown. When transistors Q_1 and Q_3 begin to conduct,

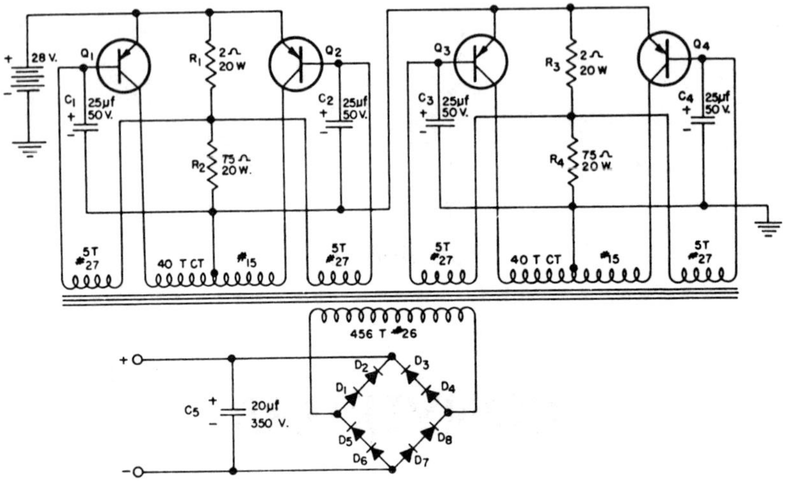

R_1, R_3 – 2Ω, 20W, wire wound resistor

R_2, R_4 – 75Ω, 20W, wire wound resistor

C_1, C_2, C_3, C_4 – 25 mfd, 50V electrolytic capacitor

C_5 – 20 mfd, 350V electrolytic capacitor

D_1, D_2, D_3, D_4, D_5, D_6, D_7, D_8 – Silicon diode (Sarkes Tarzian M500 or equivalent)

Q_1, Q_2, Q_3, Q_4 – 2N1100 Delco Hi-Power Transistor

T_1 – Transformer, core is Westinghouse RH-79 Toroid (or equivalent), primary and feedback windings are bifilar.

(A) Converter circuit.

Series-connected

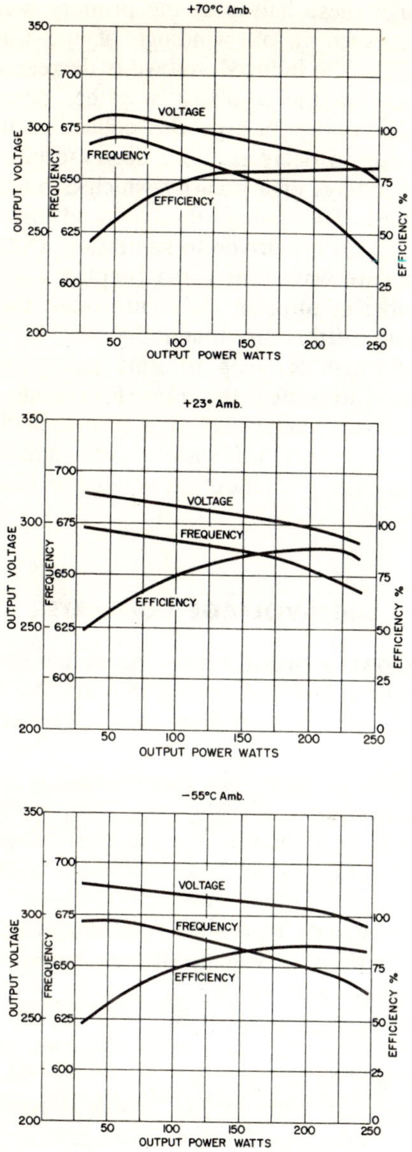

(B) Voltage, frequency, and efficiency vs ambient temperature.

Courtesy Delco Radio Div., General Motors Corp.

dc-to-dc converter.

the current through these halves of the primary windings increases, and voltage is induced in all windings of the transformer; when saturation is reached, the induced voltage is decreased to zero. When the voltage induced in the feedback windings decreases and these two transistors are not saturated, the collector current decreases, causing the core flux to decrease. This decrease induces a voltage in the transformer windings of a polarity which is opposite to the original induced voltage. The voltage thus induced biases Q_1 and Q_3 to cutoff, while Q_2 and Q_4 are driven to saturation. Output voltage from the circuit is a square wave where the frequency and amplitude depend on the number of turns in the transformer, the supply voltage, and the flux density of the core material.

Resistors R_1 through R_4 help to start oscillation, capacitors C_1 through C_4 are used to protect the transistors, while all of the diodes used in the circuit are connected in a bridge, with two diodes in each bridge leg to provide full-wave rectification. With an output power of 175 watts, there is an output voltage of 300 volts at a 23°C ambient temperature.

HIGH-VOLTAGE CONVERTER

This circuit shows a two-transistor converter that will change a 3-volt d-c input to a high-voltage a-c output. An output of up to 2100 volts ac can be obtained from this converter with this input voltage.

When switch SW1 is closed, the battery voltage appears across the primary of the transformer T1. The closed-loop path is completed by the leakage current to transistor Q2, and this voltage is then enough to drive transistor Q1 to the on condition, causing it to begin conduction.

When Q1 is turned on, it provides sufficient drive to turn on transistor Q2, and the primary voltage of transformer T1 will decrease in an exponential fashion. When the primary voltage drops below the required voltage to keep transistor Q1 on, this transistor turns off, the magnetic field of the transformer collapses, and this action induces a 60-volt field in the primary of T1. At the same time, this produces the necessary high voltage, which is about 2000 volts, in the secondary. At this time Q1 turns on again so that the cycle is repeated. Diode D1 is a protective diode; it is in the circuit so that the voltage of the primary is not impressed across the base-to-emitter junction of transistor Q1.

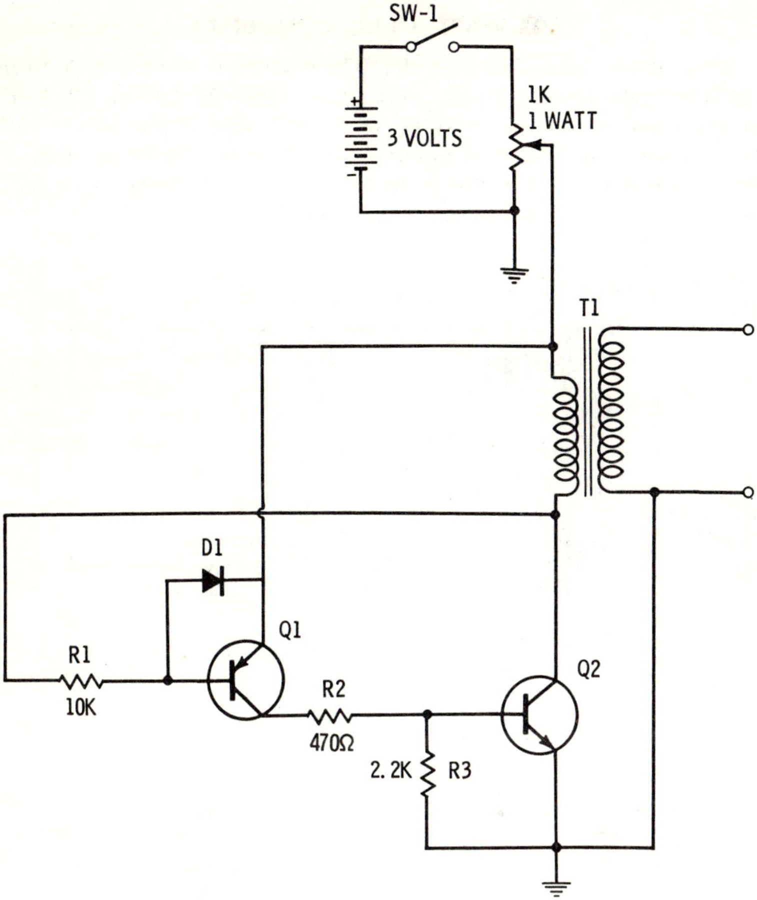

NOTES:

D1: GD-426

Q1: GT-2904

Q2: GT-2903

ALL RESISTORS ARE ± 10% 1/2 WATT

T1: 50:1 TURNS RATIO

D C RESISTANCE: PRI = 3Ω SEC. =3000Ω

MICROTRAN TYPE M8050

Courtesy General Instrument Corp.

High-voltage converter.

225-WATT 10-KC CONVERTER

This circuit shows two type 2N1936 transistors used in a dc-to-dc converter operating at 10 kc. This circuit operates with an efficiency of 80% and an output ripple of less than 1 volt; it provides a 150-volt output from a 28-volt input source. This is a switching type of dc-to-dc converter. The power transistors and the diodes must have adequate heat sinks for proper operation.

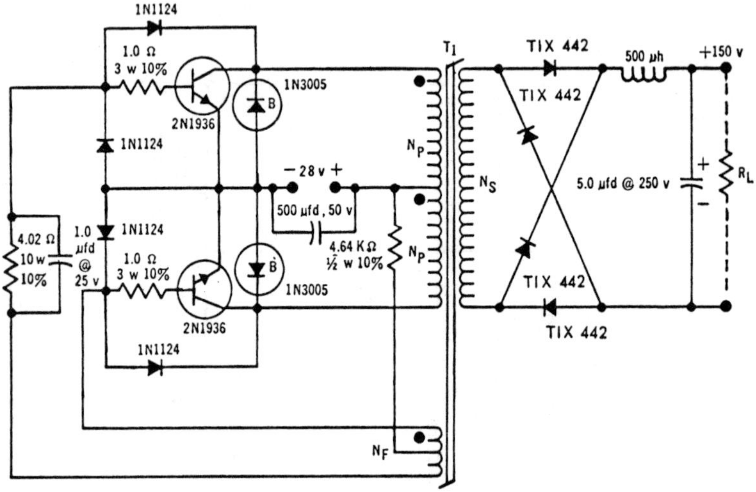

NOTE: Minimun usable h_{FE} = 10 at 10 amps, V_{CE} = 3v. Both transistors use heat sinks, $\theta_{CHS} + \theta_{HS\text{-}A} \leqslant$ 2 C/w each. All diodes must have adequate heat sinks.

Transformer data:
Core = Arnold 2T-6847 - S1 or equivalent
N_p = 15 turns #13 heavy Formvar
N_F = 6 turns, tapped at 3 turns #18
N_S = 87 turns #18

Courtesy Texas Instruments, Inc.

225-watt 10-kc converter.

DC-TO-DC CONVERTER

This is a 120-watt dc-to-dc converter using two type 2N1722 transistors in the switching circuit. Four 1N1096 silicon-rectifier diodes are used in the bridge-circuit output to provide 400 volts at 120 watts. This converter circuit has an input current of 5 amperes and an output ripple of 0.6 volt maximum. The overall circuit efficiency is 85%.

This is a self-starting switching circuit where the conducting transistor builds up a magnetic field until saturation is reached, and the collapsing magnetic field turns the other transistor on. Both transistors are mounted on a dual-transistor heat sink.

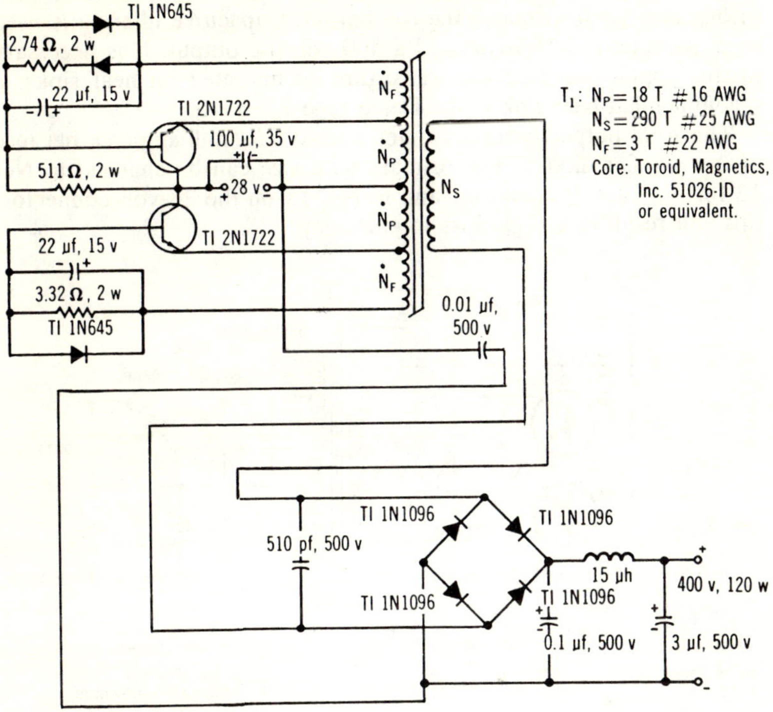

NOTES: 1. All Resistance Values in ohms, 5% Tolerance.
2. All Resistor Wattage Ratings at 125°C Ambient.
3. Capacitor Voltage Ratings at 125°C Ambient.
4. Both Transistors on Same Heat Sink, $\theta_{C\text{-}HS} + \theta_{HS\text{-}A}$ 4 C°/w each.

Courtesy Texas Instruments, Inc.

Dc-to-dc converter.

145/225-WATT CONVERTER

This circuit shows a two-transistor power converter with an output of 145 watts with an input of 14 volts, and a power output of 225 watts with an input of 28 volts. Information on both of these applications is shown in the parts list.

This circuit, like many other converters, is essentially a power flip-flop where one transistor is conducting while the other is cut off. The result is a square-wave output, and as shown in the circuit, operation is at about 800 cycles. Resistors R1 and R2 are used to start oscillation by biasing the transistors, while the four diodes form a bridge rectifier to produce the d-c output. Capacitive filtering is used to provide both a 150-volt and a 300-volt d-c output. It is necessary in this application that the transistors be mounted on heat sinks; a recommended heat sink is the Delco type 7276040.

The total output current cannot exceed 750 milliamperes because of the rectifier diodes. The primary wire size can be changed to No. 13 for the 14-volt connector and to No. 15 on the 28-volt connector; this will result in a slight loss in efficiency.

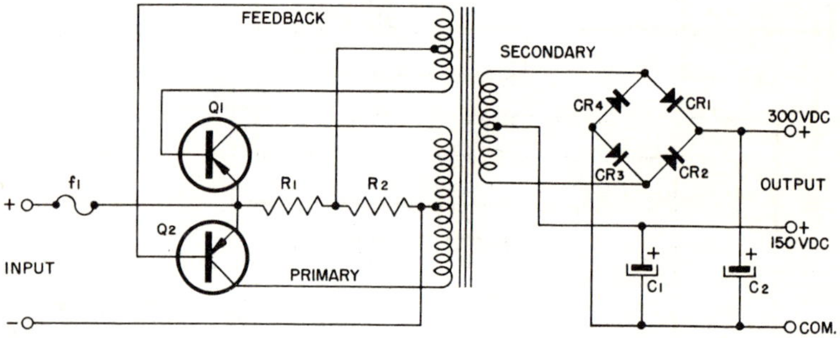

Part	14V Converter	28V Converter
Q_1, Q_2	Delco 2N174	Delco 2N1412
R_1	1.5 ohm 10W wirewound	3 ohm 5W wirewound
R_2	40 ohm 10W wirewound	150 ohm 10W
$CR_{1,2,3,4}$	RCA 1N3195	RCA 1N3195
C_1	10 uf 200Vdc	10 uf 200Vdc
C_2	30 uf 350Vdc	30 uf 350Vdc
f_1	20 amp fuse	15 amp fuse
Toroid Core	Magnetics Inc. Butler, Pa. 50001-2A or 51001-2A	Magnetics Inc. Butler, Pa. 50001-2A or 51001-2A

Transformer Winding	14V	28V
Primary*	40t Tap at 20t No. 10 A.W.G.	80t Tap at 40t No. 12 A.W.G.
Feedback*	12t Tap at 6t No. 18 A.W.G.	12t Tap at 6t No. 18 A.W.G.
Secondary	448t Tap at 224 No. 23 A.W.G.	440t Tap at 220t No. 23 A.W.G.

* Bifilar wound

Courtesy Delco Radio Div., General Motors Corp.

145/225-watt converter.

DC-TO-DC CONVERTER

This is a common-collector circuit operating as a dc-to-dc converter to produce 100 watts at 400 cycles into the rectifier. Transistors type 2N443 are used with a 12-volt source, producing 500 volts of output to the load. Type SKM500 diodes are used as the rectifiers.

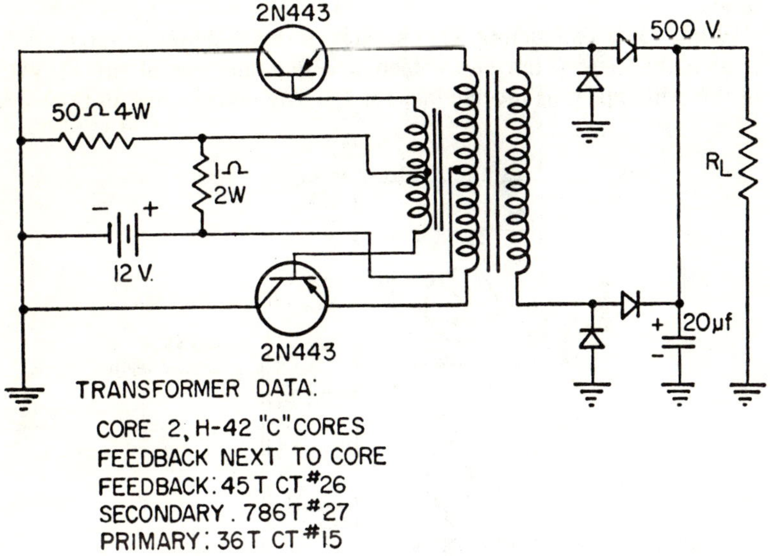

Courtesy Delco Radio Div., General Motors Corp.

Dc-to-dc converter.

6-TO-12 VOLT DC-TO-DC CONVERTER WITHOUT RECTIFIERS

This circuit is a dc-to-dc converter that provides a 12-volt d-c output using a 6-volt d-c source; the frequency of operation is 300 cycles.

A pair of 2N443 transistors is used as shown. The necessary voltage amplification or multiplication is obtained by means of adding the supply voltage to the feedback voltage. In operation, assume that transistor Q_1 is turning on. Voltages will be generated with the polarities as shown in the schematic; thus a negative voltage is applied to the base of transistor Q_1. Because of these polarities, there is a positive feedback, and transistor Q_1 will conduct until the transformer is saturated. At this point there is about 6.8 volts across the upper half of the primary winding, which develops 8 volts in each half of the feedback winding. Voltage in the lower half of the feedback winding adds to the supply voltage to provide approximately 14 volts across the load.

Current path, as shown, is from the lower end of the feedback winding, through the emitter-to-base diode, to the positive end of the input voltage. Voltage in the feedback winding is thus added to the 6-volt source, and there is about 0.6- to 1-volt drop across this diode.

While Q_1 is conducting, Q_2 operates in this fashion; about 16 volts is available across the entire feedback winding and about 15 volts of this will appear as reverse bias for Q_2. The reverse bias on Q_2 is due

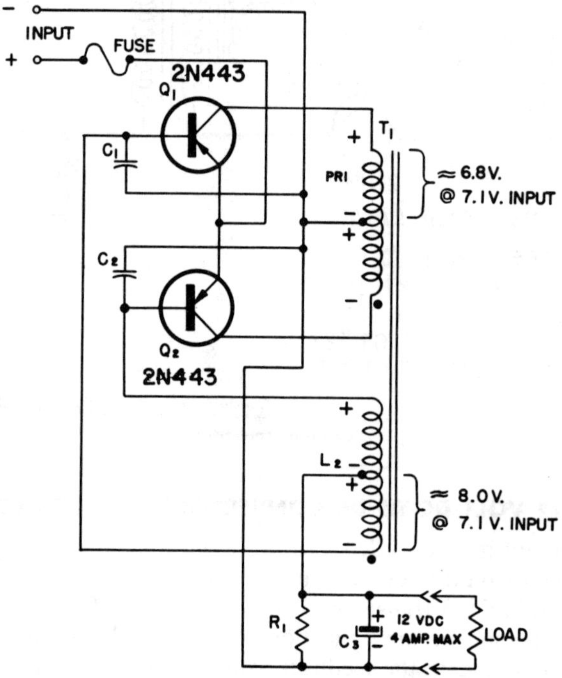

Q_1, Q_2—Delco 2N443
C_1, C_2—.5 mfd, 100V paper
C_3—1500 mfd, 15V
R_1—470 ohm, 1 watt

Fuse—10 amp. Slo-Blo
T_1—L_1 (Primary) 48 Turns #17, tap @ 24 t
L_2 (Feedback) 56 Turns #17, tap @ 28 t
Core: H-3 Hypersil Type "C" Core

Courtesy Delco Radio Div., General Motors Corp.

6-to-12 volt dc-to-dc converter.

to the forward conducting voltage of the emitter-to-base diode of Q_1. When the transformer saturates, the magnetic field collapses and voltages of a polarity opposite to those shown are generated; Q_1 is now cut off, and Q_2 begins to conduct.

In this circuit the emitter-to-base diodes can be considered as rectifying diodes in a full-wave circuit, so the d-c output is taken from the center tap of the feedback winding. R_1 is used to assist in the starting of oscillations, C_1 and C_2 are used for suppression of the spikes, and C_3 is the filter capacitor.

Both transistors should be mounted on one Delco 7281352 heat sink.

DC-TO-DC CONVERTER

This circuit uses type 2N442 transistors and type SKM500 diodes. This common-base circuit produces 100 watts of output power at 800 cycles into the rectifier. The original battery source is 12 volts, and the output is either 250 volts or 500 volts, as shown.

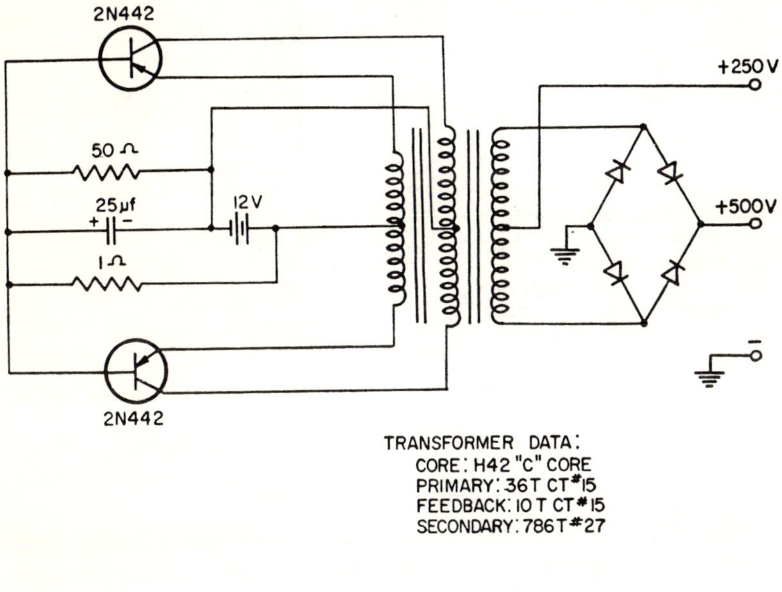

Courtesy Delco Radio Div., General Motors Corp.

Dc-to-dc converter.

DC-TO-DC CONVERTER (30-VOLT TO 480-VOLT)

This is a power converter for changing a low d-c voltage to a high d-c voltage; the input is 30 volts, and the output is 480 volts at 250 ma.

Four power transistors are used. Each transistor has a feedback winding, with a 2-ohm resistor in series, from its emitter to base. One transformer with four feedback windings is used, one winding for each power transistor. The primary winding has 200 turns and the output winding has 3470 turns. A bridge rectifier, with two diodes in series for each leg, is used to rectify the ac across the output winding to produce the high d-c voltage.

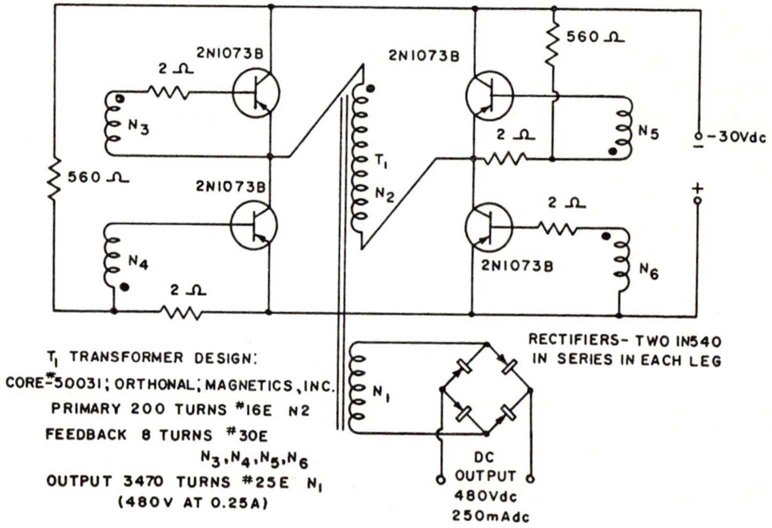

Courtesy CBS Electronics Div., CBS Inc.

Dc-to-dc converter.

DC-TO-DC CONVERTERS

Two different dc-to-dc converters are shown using 2N256 transistors. The maximum switching ratings for these transistors are 2 amperes for each collector and 100 ma for each base. Meters are shown so that these readings can be monitored, since various input voltages can be used. The output voltages obtained with these two

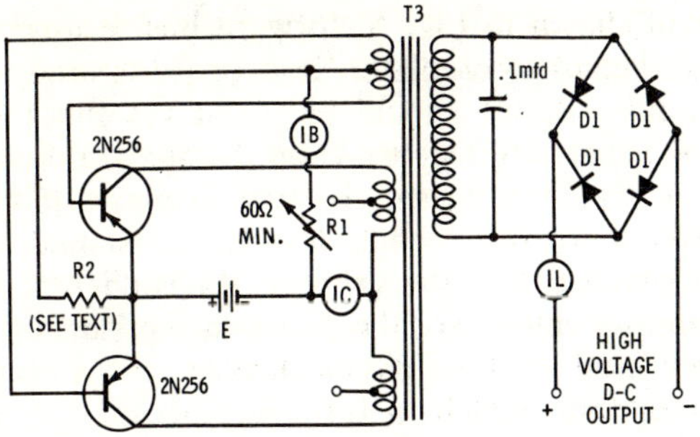

(A) Resistor used for regulation.

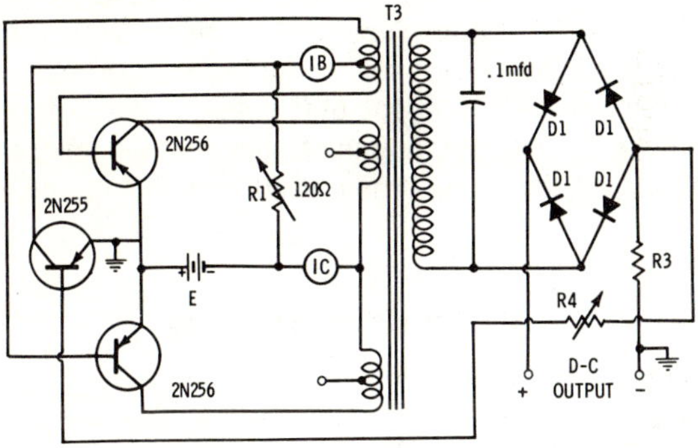

(B) Transistor provides self-regulation.

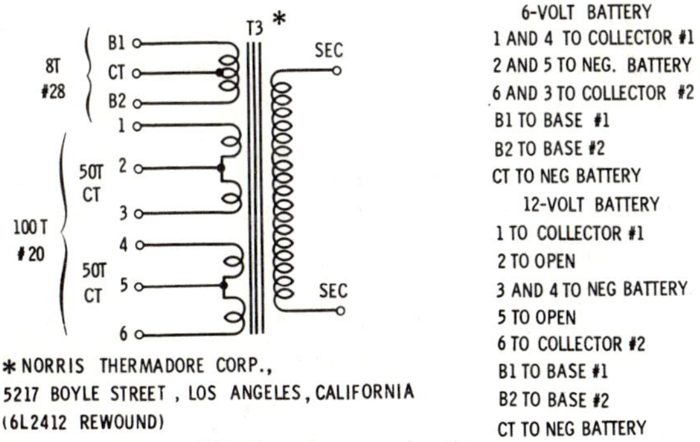

(C) Transformer details.

Courtesy CBS Electronics Div., CBS Inc.

Dc-to-dc converters.

circuits depend on the input voltage used; a range of 2- to 350-volts d-c output can be obtained for input voltages of 6 to 12 volts.

In the circuit shown in Fig. A, forward bias is used to make sure that the circuit begins to oscillate. Resistor R2 is used for regulation. The value of this resistor should be about ten times the resistance of R1, and the value of R1 depends on the applied d-c input voltage.

In Fig. B a 2N255 transistor is used in place of R2 to provide for self-regulation. In this circuit, resistor R3 should be about 5% of the load resistance, and the value of R4 is chosen to provide for the required output voltage for the particular application. The rectifier diodes that are used in either circuit depend on the voltage available across the secondary, which in turn depends on the input voltage. Typical rectifiers are the 1N1764 which has a voltage rating of 500 volts at an average forward current of 0.5 ampere, and the 1N3256 which has a maximum voltage of 800 volts and has the same current rating as the 1N1764. Other suggested rectifiers with a forward current rating of 0.75 ampere are the 1N3253 for 200 volts, the 1N3254 for 400 volts, or the 1N3255 for 600 volts.

Fig. C shows the suggested transformer as used in these two converters. Details of the transformer used and the model number are shown.

DC-TO-DC CONVERTER

This circuit shows a dc-to-dc converter with suggested transistor types 2N1358, 2N1100, 2N443, and 2N174. The suggested rectifiers are types 1N1763, 1N1764, 1N3196, and 1N3256.

Various types of transformers can provide different types of voltage output. For Triad transformers TY78 through TY84 the input is 12.6 volts d-c. Two voltage outputs can be obtained, depending upon whether the circuit is connected as a full-wave bridge as shown, in which case the higher output voltage is produced, or whether the circuit is connected as a full-wave center tap, in which case the lower output voltage is available. For the TY78 transformer the output voltage is 250 volts or 125 volts, at 100 ma. For the TY79 transformer the output is 300 or 150 volts, at 200 ma. For the TY80 transformer the output is 325 or 162.5 volts, at 150 ma. For the TY81 transformer the output is 375 or 187.5 volts, at 200 ma. For the TY82 transformer the output is 450 or 225 volts, at 150 ma. For the TY83 transformer the output is 500 volts or 250 volts, at 250 ma. For the TY84 transformer the output is 600 or 300 volts, at 200 ma.

Parts List

	$R_1 = R_2$	$R_3 = R_4$	C_1	C_2	C_3	C_4	C_5
TY-78	500 ohm, 1 watt	16 ohm, ½ watt	250 mfd, 25V	1-4 mfd, 450V	.001 mfd, 1000V	2 mfd, 25V	2 mfd, 25V
TY-79	250 ohm, 2 watt	7 ohm, 1 watt	250 mfd, 25V	1-4 mfd, 450V	.001 mfd, 1000V	5 mfd, 25V	5 mfd, 25V
TY-80	300 ohm, 2 watt	8 ohm, ½ watt	250 mfd, 25V	1-4 mfd, 450V	.001 mfd, 1000V	5 mfd, 25V	5 mfd, 25V
TY-81	300 ohm, 2 watt	10 ohm, 1 watt	250 mfd, 25V	1-4 mfd, 500V	.005 mfd, 1600V	2 mfd, 25V	2 mfd, 25V
TY-82	300 ohm, 5 watt	7 ohm, 1 watt	250 mfd, 25V	1 mfd, 1000V	.005 mfd, 1600V	2 mfd, 25V	2 mfd, 25V
TY-83	100 ohm, 5 watt	2 ohm, 2 watt	250 mfd, 25V	1 mfd, 1000V	.001 mfd, 1600V	2 mfd, 25V	2 mfd, 25V
TY-84	150 ohm, 5 watt	2.5 ohm, 2 watt	250 mfd, 25V	1 mfd, 1000V	.001 mfd, 1600V	2 mfd, 25V	2 mfd, 25V
TY-89	5K, 2 watt	50 ohm, 1 watt	250 mfd, 50V	1-4 mfd, 450V	.005 mfd, 1000V	2 mfd, 50V	2 mfd, 50V
TY-90	2.5K, 5 watt	25 ohm, 2 watt	250 mfd, 50V	1-4 mfd, 450V	.005 mfd, 1000V	2 mfd, 50V	2 mfd, 50V
TY-91	2.5K, 1 watt	34 ohm, 2 watt	250 mfd, 50V	1-4 mfd, 500V	.005 mfd, 1600V	2 mfd, 50V	2 mfd, 50V
TY-92	3K, 5 watt	30 ohm, 1 watt	250 mfd, 50V	1-4 mfd, 1000V	.005 mfd, 1600V	2 mfd, 50V	2 mfd, 50V
TY-93	1250 ohm, 5 watt	30 ohm, 1 watt	250 mfd, 50V	1-4 mfd, 1000V	.001 mfd, 1600V	2 mfd, 50V	2 mfd, 50V
TY-94	2.5K, 5 watt	20 ohm, 1 watt	250 mfd, 50V	1-4 mfd, 1000V	.004 mfd, 1600V	2 mfd, 50V	2 mfd, 50V

For a 28-volt d-c input other transformer types are available as shown. These are Triad types TY89 through TY94. For transformer type TY89 the output is 300 or 150 volts, at 100 ma. For type TY90 the output is 325 or 162.5 volts, at 200 ma. For type TY91 the output is 375 or 187.5 volts dc, at 200 ma. For type TY92 the output is 450 or 225 volts, at 200 ma. For type TY93 the output is 500 or 250 volts, at 250 ma. For type TY94 the output is 600 or 300 volts dc, at 200 ma.

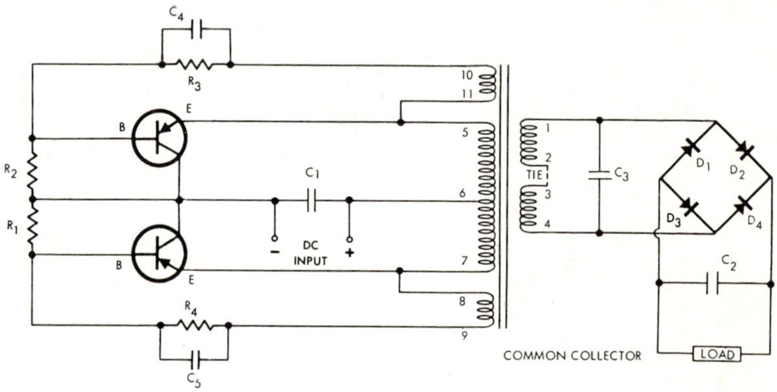

Courtesy Triad Transformer Corp.

Dc-to-dc converter.

DC-TO-DC CONVERTER

This circuit shows a common-emitter dc-to-dc converter with suggested transistor types 2N173, 2N174, 2N443, 2N1100, or 2N1358. Rectifier types that are suggested are 1N1764 and 1N3196.

Two different voltage outputs are available, depending on whether the circuit is connected in a full-wave bridge configuration where the higher output is available, or connected as a full-wave center tap where the lower voltage output is obtained.

Triad transformers type TY78 through TY86 are available for use with a 12.6-volt dc input. Various voltage outputs are available with the different transformers. For transformer TY-78, 250 or 125 volts at 100 ma is available. For type TY79, 300 or 150 volts dc at 200 ma is available. For type TY80, 325 or 162.5 volts at 150 ma is available. For type TY-81, 375 or 187.5 volts dc at 200 ma can be obtained. For type TY-82, 450 or 225 volts at 150 ma can be obtained.

Parts List

	Input	Output	R_1	R_2	C_1	C_2	C_3	C_4
TY-78	12.6	250/125 @ 100 ma	400 ohm, 5 watt	20 ohm, 1 watt	250 mfd, 25V	1-4 mfd, 450V	.001 mfd, 1000V	2 mfd, 25V
TY-79	12.6	300/150 @ 200 ma	160 ohm, 5 watt	7.5 ohm, 1 watt	250 mfd, 25V	1-4 mfd, 450V	.001 mfd, 1000V	5 mfd, 25V
TY-80	12.6	325/162.5 @ 150 ma	280 ohm, 2 watt	10 ohm, 1 watt	250 mfd, 25V	1-4 mfd, 450V	.001 mfd, 1000V	5 mfd, 25V
TY-81	12.6	375/187.5 @ 200 ma	500 ohm, 2 watt	25 ohm, 2 watt	250 mfd, 25V	1-4 mfd, 500V	.005 mfd, 1600V	2 mfd, 25V
TY-82	12.6	450/225 @ 150 ma	400 ohm, 5 watt	10 ohm, 2 watt	250 mfd, 25V	1 mfd, 1000V	.005 mfd, 1600V	2 mfd, 25V
TY-83	12.6	500/250 @ 250 ma	75 ohm, 5 watt	5 ohm, 5 watt	250 mfd, 25V	1 mfd, 1000V	.001 mfd, 1600V	2 mfd, 25V
TY-84	12.6	600/300 @ 200 ma	75 ohm, 5 watt	5 ohm, 5 watt	250 mfd, 25V	1 mfd, 1000V	.001 mfd, 1600V	2 mfd, 25V
TY-85	12.6	600/300 @ 350 ma	50 ohm, 10 watt	2 ohm, 10 watt	250 mfd, 50V	1-4 mfd, 1000V	.004 mfd, 1600V	2 mfd, 50V
TY-86	12.6	425/212.5 @ 350	50 ohm, 10 watt	5 ohm, 10 watt	250 mfd, 25V	1-4 mfd, 1000V	.004 mfd, 1600V	2 mfd, 50V
TY-89	28	300/150 @ 100 ma	2.5K, 10 watt	50 ohm, 1 watt	250 mfd, 50V	1-4 mfd, 450V	.005 mfd, 1000V	2 mfd, 50V
TY-90	28	325/162.5 @ 200 ma	1K, 2 watt	30 ohm, 2 watt	250 mfd, 50V	1-4 mfd, 450V	.005 mfd, 1000V	2 mfd, 50V
TY-91	28	375/187.5 @ 200 ma	5K, 1 watt	40 ohm, 2 watt	250 mfd, 50V	1-4 mfd, 500V	.005 mfd, 1600V	2 mfd, 50V
TY-92	28	450/225 @ 200 ma	4K, 2 watt	50 ohm, 2 watt	250 mfd, 50V	1-4 mfd, 1000V	.005 mfd, 1600V	2 mfd, 50V
TY-93	28	500/250 @ 250 ma	3K, 5 watt	45 ohm, 2 watt	250 mfd, 50V	1-4 mfd, 1000V	.001 mfd, 1600V	2 mfd, 50V
TY-94	28	600/300 @ 200 ma	2.5K, 5 watt	35 ohm, 2 watt	250 mfd, 50V	1-4 mfd, 1000V	.004 mfd, 1600V	2 mfd, 50V
TY-99	6.3	300/150 @ 100 ma	300 ohm, 2 watt	20 ohm, 5 watt	250 mfd, 25V	1-4 mfd, 450V	.001 mfd, 1000V	2 mfd, 25V
TY-100	6.3	325/162.5 @ 150 ma	100 ohm, 2 watt	5 ohm, 5 watt	250 mfd, 25V	1-4 mfd, 450V	.001 mfd, 1000V	2 mfd, 25V
TY-101	6.3	375/187.5 @ 200 ma	55 ohm, 5 watt	2 ohm, 5 watt	250 mfd, 25V	1-4 mfd, 500V	.001 mfd, 1600V	2 mfd, 25V
TY-102	6.3	450/225 @ 150 ma	38 ohm, 5 watt	7 ohm, 5 watt	250 mfd, 25V	1-4 mfd, 1000V	.005 mfd, 1600V	5 mfd, 25V

For type TY-83, 500 or 250 volts at 250 ma can be obtained. For type TY-84, 600 or 300 volts at 200 ma can be obtained. For type TY-85, 600 or 300 volts at 350 ma can be obtained. For type TY-86, 425 or 212.5 volts dc at 350 ma can be obtained.

For a 28-volt d-c input various other transformers are available, as indicated. For type TY-89, 300 or 150 volts dc at 100 ma can be obtained. For type TY-90, 325 or 162.5 volts dc at 200 ma can be obtained. For type TY-91, 375 or 187.5 volts at 200 ma can be obtained. For type TY-92, 450 or 225 volts at 200 ma can be obtained. For type TY-93, 500 or 250 volts dc at 250 ma can be obtained. For type TY-94, 600 or 300 volts at 200 ma can be obtained.

There are also other voltage outputs available by using a 6.3-volt input in place of the two voltage inputs just given. Using a transformer type TY-99, an output of 300 or 150 volts at 100 ma can be obtained. Using transformer type TY-100, an output of 325 or 162.5 volts at 150 ma can be obtained. Using transformer type TY-101, an output of 375 or 187.5 volts at 200 ma can be obtained. Using transformer type TY-102, an output of 450 or 225 volts at 150 ma can be obtained.

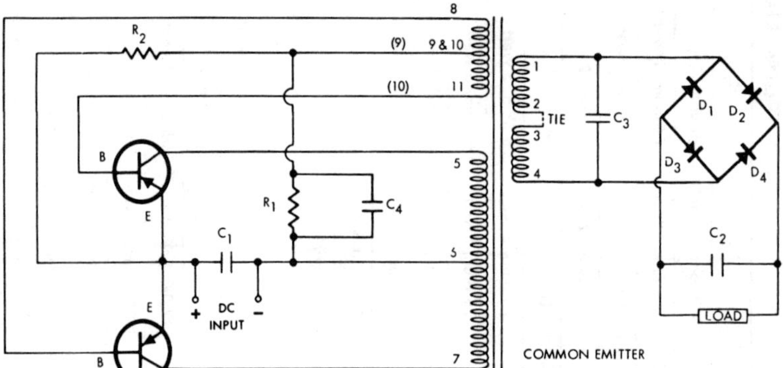

Courtesy Triad Transformer Corp.

Common-emitter dc-to-dc converter.

DC-TO-DC CONVERTER

This circuit shows several possible dc-to-dc converter arrangements using transistor types 2N173, 2N174, or 2N443, and rectifier types 1N1763, 1N1764, or 1N3256. In each case the input is 12.6 volts dc.

In this common-emitter circuit three different Triad transformers are suggested for three different output ranges. For transformer type TY68S the output is 250 volts using a full-wave bridge rectifier, and

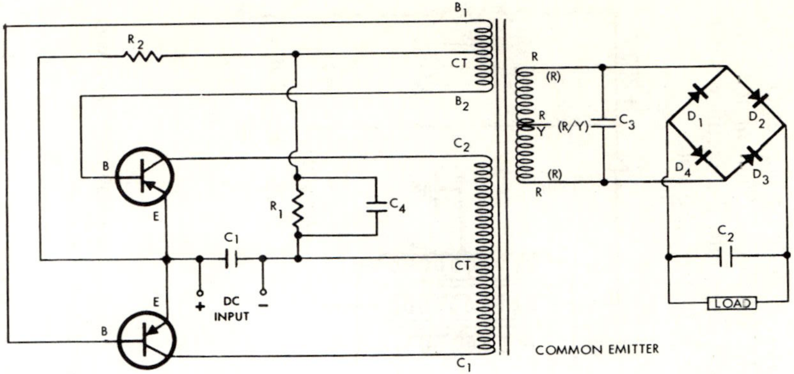

	Input	Output	R_1	R_2	C_1	C_2	C_3	C_4
TY-68S	12	250/125 @ 65 ma	60 ohm, 5 watt	600 ohm, 1 watt	250 mfd, 25V	1-4 mfd, 450V	.005 mfd, 1000V	25 mfd, 25V
TY-70S	12	325/162.5 @ 150 ma	91 ohm, 1 watt	5.6 ohm, 1 watt	250 mfd, 25V	1-4 mfd, 450V	.005 mfd, 1000V	25 mfd, 25V
TY-71S	12	375/187.5 @ 200 ma	36 ohm, 2 watt	2.7 ohm, 4 watt	250 mfd, 25V	1-4 mfd, 500V	.005 mfd, 1600V	10 mfd, 25V

Courtesy Triad Transformer Corp.

Dc-to-dc converter.

125 volts using a full-wave center tap; in both cases the current maximum is 65 ma. With transformer type TY70S the output is either 325 or 162.5 volts dc at 150 ma. Using transformer type TY-71S the output is 375 or 187.5 volts dc at 200 ma.

DC-TO-DC CONVERTER

This is a common-emitter type of dc-to-dc converter using type 2N441 transistors and type SKM500 diodes. The input to the converter is a 6-volt battery, and the input into the rectifier is 50 watts at 400 cycles. The circuit has an output of 250 volts.

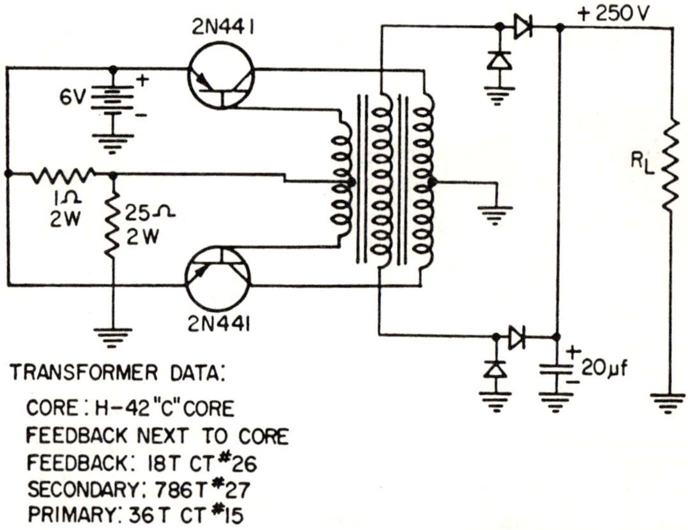

Courtesy Delco Radio Div., General Motors Corp.

Common-emitter dc-to-dc converter.

DC-TO-DC CONVERTER

This is a dc-to-dc converter or voltage multiplier that can step up a small d-c voltage to a large d-c voltage. Since various voltage step-ups are possible, the circuit as shown has three meters, one for measuring the base current, one for the emitter current, and one for the output current.

For the transistors shown, the maximum switching rating is 2 amperes for each collector and 100 ma for each base. A control is adjusted to provide currents within these ranges for the specific voltage used for the input.

Suggested transformers for a voltage step-up from 6 volts to 250 volts are shown; other transformers can be used provided that the same current ratio is maintained.

T1 turns ratio is $\frac{E_{OUT}}{2} \div E_{IN}$; for an E_{OUT} of 250 volts dc and an E_{IN} of 6 volts dc, a filament transformer (115 volts/6 volts C.T.) can be used if its power rating is at least equal to the power output. T2 is a tube-to-voice-coil transformer of 2000/8-4 ohms impedance. D-1 can be a 1N1763 or 1N443B.

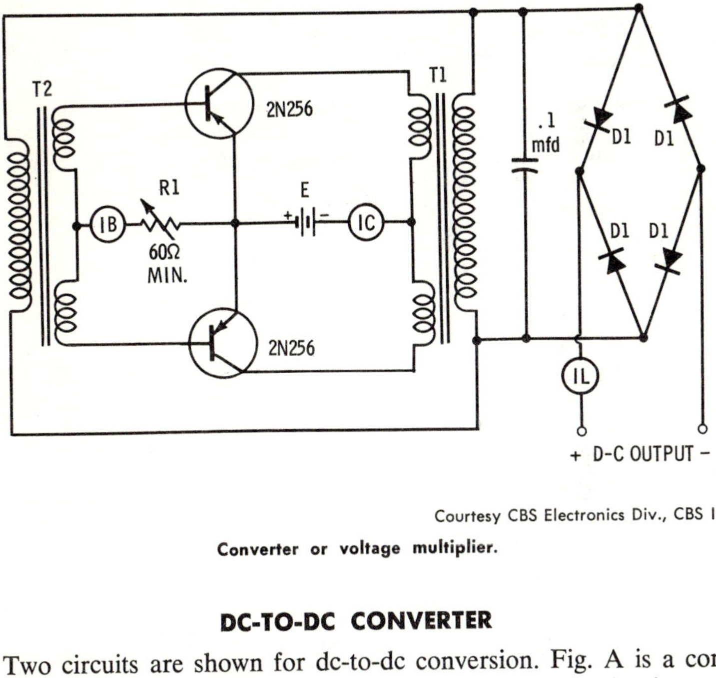

Courtesy CBS Electronics Div., CBS Inc.

Converter or voltage multiplier.

DC-TO-DC CONVERTER

Two circuits are shown for dc-to-dc conversion. Fig. A is a common-emitter circuit, while Fig. B is a common-collector circuit.

Figs. A and B both use Triad TY88 transformers. The supply voltage is 28 volts dc, and if it is connected as a full-wave bridge circuit, the output voltage is 250 volts dc. If the input is connected as a full-wave center-tap rectifier, the output voltage would be half this value, or 125 volts dc. The maximum current drain is 80 ma. The suggested transistors are type 2N173, 2N174, 2N443, or 2N1099. Suggested rectifiers are 1N1763, 1N1764, 1N3196, or 1N3256.

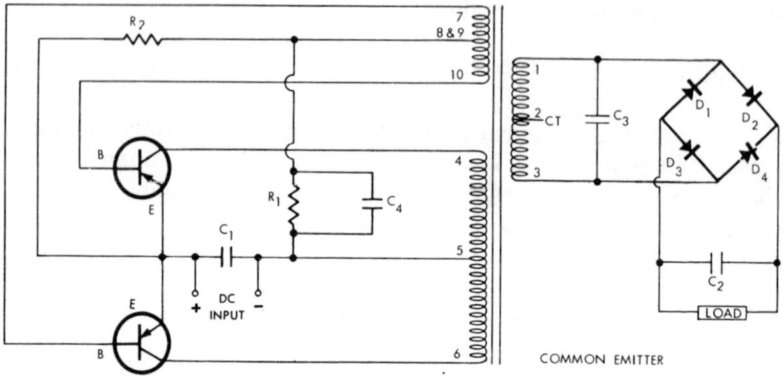

R_1—5K, 1 watt
R_2—60 ohm, 1 watt
C_1—250 mfd, 50V
C_2—1-4 mfd, 450V
C_3—.001 mfd, 1000V
C_4—2 mfd, 50V

(A) Common-emitter circuit.

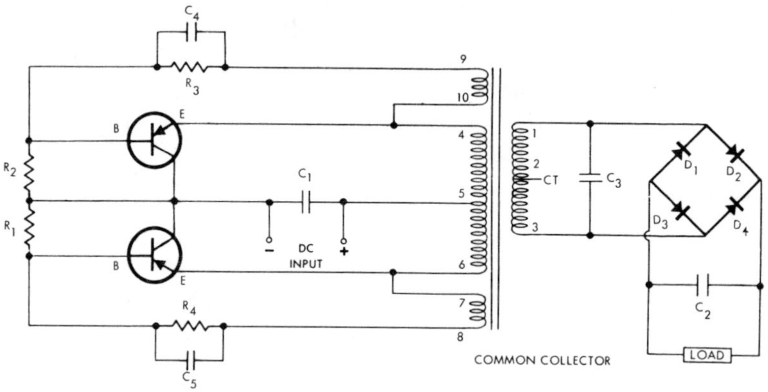

R_1, R_2—5K, 1 watt
R_3, R_4—200 ohm, 1 watt
C_1—250 mfd, 50V
C_2—1-4 mfd, 450V
C_3—.001 mfd, 1000V
C_4—2 mfd, 50V
C_5—2 mfd, 50V

(B) Common-collector circuit.

Courtesy Triad Transformer Corp.

Dc-to-dc converter circuits.

DC-TO-DC CONVERTER

This is a common-emitter type of dc-to-dc converter using 2N442 transistors. This circuit produces 100 watts of 800-cycle power into the rectifier, which uses four type SKM500 diodes. The input power supply is 12 volts, and the output of the converter is at two levels: +250 volts and +500 volts.

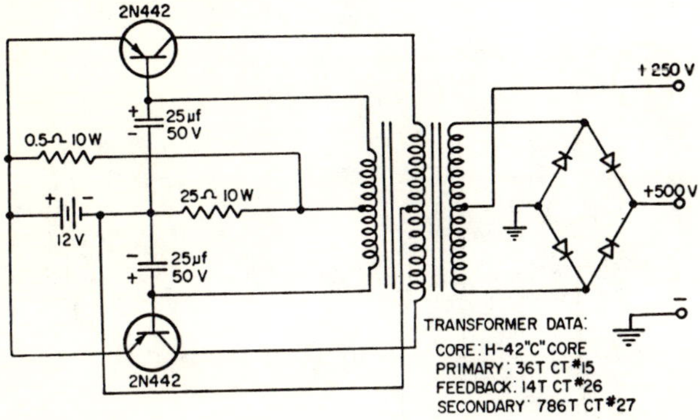

Courtesy Delco Radio Div., General Motors Corp.

Dc-to-dc converter.

DC-TO-DC CONVERTER

This circuit shows two 2N441 transistors operating as a common-collector dc-to-dc converter from 6 volts to 250 volts. All diodes are type SKM500, and the circuit provides 50 watts output at 400 cycles into the rectifier circuit.

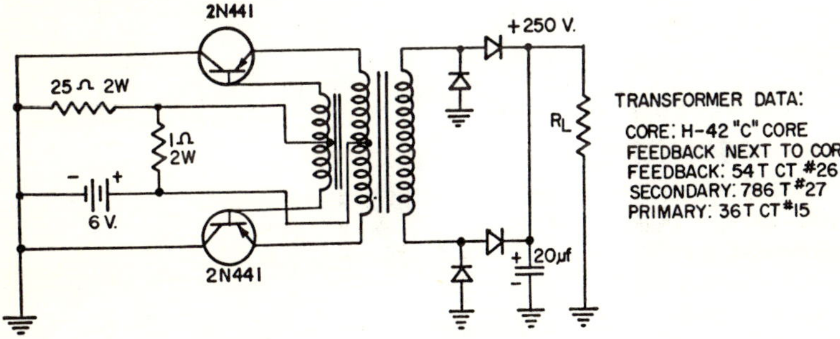

Courtesy Delco Radio Div., General Motors Corp.

Dc-to-dc converter.

Section 4

INVERTERS

DC-TO-AC INVERTER

This circuit shows a dc-to-ac inverter using type 2N1073B transistors with fast switching times. The switching frequency of this circuit is about 2500 cycles; it is possible to add other secondary windings on the core or to add a high-voltage secondary if desired. The output from this circuit can be fed into a half-wave or bridge rectifier in order to produce d-c output from the 150-volt peak output at 120 watts.

Assume that Q_1 starts conducting because of the negative-feedback voltage. The return path for the feedback voltage is through resistor R_1 and through diode CR_2. There is a voltage drop across CR_2 when it is conducting, and this voltage is of such polarity that a reverse bias is fed to the emitter-to-base diode of Q_2. This means that transistor Q_2 is cut off while transistor Q_1 is going into saturation. When the core saturates, the magnetic field begins to collapse and the polarity of the feedback voltage reverses. Q_2 now starts to conduct, and Q_1 is cut off. Now the feedback return is through diode CR_1 and the same action occurs as when Q_1 was conducting. The voltage drop across CR_1 will assure that Q_1 is reverse biased, and therefore cut off.

Capacitors C_1 and C_2 are used to remove the spikes from the collector-to-emitter waveform. The reverse bias on the emitter diodes never is greater than the voltage across CR_1 and CR_2, thus the emitter diodes are protected.

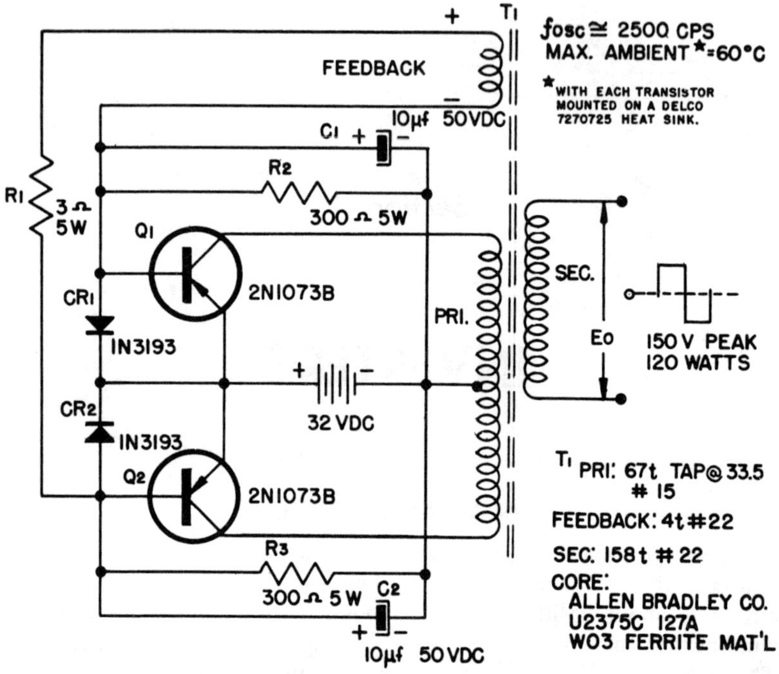

Courtesy Delco Radio Div., General Motors Corp.

Dc-to-ac inverter.

HIGH-CURRENT INVERTER

The circuit in Fig. A shows a pair of 2N2728 power transistors operating in a low-voltage high-current inverter circuit. This inverter will change a 2-volt d-c source to a higher-voltage square wave. The circuit operates in a common-emitter arrangement. The secondary windings of the driver transformers are split to allow each transistor to be driven separately. Resistors R1 through R4 are used to provide a voltage divider and starting bias for the circuit. Fig. B shows the transformer specifications.

This inverter operates at about 1000 cycles per second with an efficiency of about 70%; the power output is 60 watts.

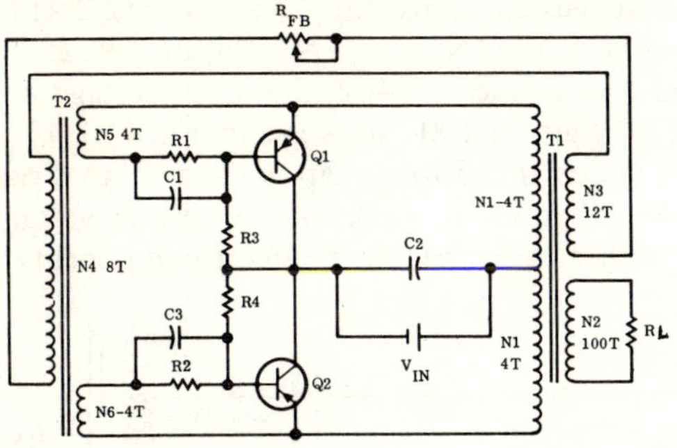

(A) Circuit diagram.

Driver Transformer PX-2126

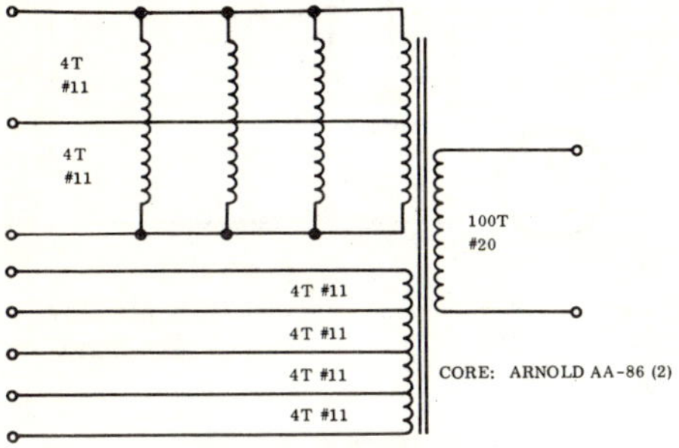

Output Transfomer PX-2127

(B) Transformer specifications.

R_1—0.75 ohm, 5 watt
R_2—0.75 ohm, 5 watt
R_3—7.5 ohm, 5 watt
R_4—7.5 ohm, 5 watt
R_{FB}—1 ohm, 5 watt
C_1—20 mfd, 6V
C_2—10,000 mfd, 6V
C_3—20 mfd, 6V
T_1—Phoenix Transformer PX 2127
T_2—Phoenix Transformer PX 2126
Q_1, Q_2—2N2728
V_{in}—2V, 50A

Courtesy Motorola Semiconductor Products

High-current inverter.

DC-TO-AC INVERTER

This circuit shows an inverter with several voltage inputs and outputs. The suggested transistors are 2N173, 2N174, or 2N443. Suggested rectifiers are 1N1763, 1N3196, or 1N3256.

With Triad transformer TY462, the input voltage is 12.6 volts dc; the output of 60 watts at 400 cycles is either at 110, 115, or 125 volts, depending on which transformer tap is used. With Triad transformer TY468, a 28-volt input is used, and the same voltage outputs are available. In either case, the other circuit components are the same.

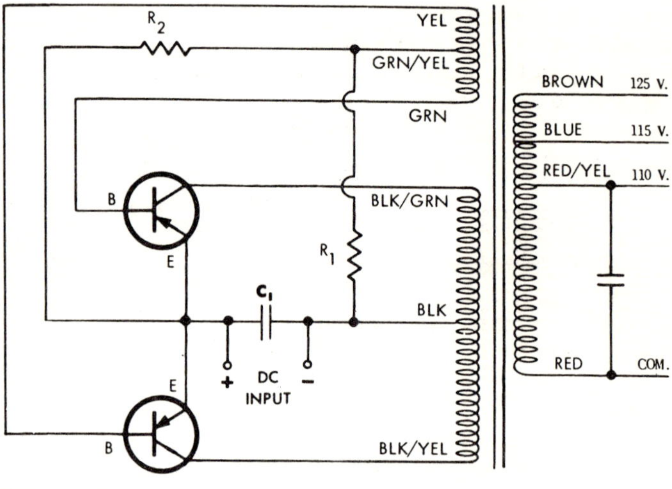

R_1—500 ohm, 10 watt
R_2—50 ohm, 1 watt

C_1—250 mfd, 50V
C_2—1 mfd, 400V

Courtesy Triad Transformer Corp.

Dc-to-ac inverter.

2-VOLT DC TO 120-VOLT AC INVERTER

This circuit shows an unusual inverter which produces 120 volts ac from a 2-volt d-c input. Transistors type 2N1518, 2N1519, 2N1520, 2N1521, 2N1522, and 2N1523 can be used in this circuit since these transistors can switch anywhere from 25 to 50 amperes.

If the 2-volt d-c source has a current capability of 25 to 50 amperes, this circuit can deliver 30 to 50 watts of 120-cycle square-wave a-c output.

Only one bias resistor is used and a separate feedback transformer is not required. The circuit shows a 25-ampere 2-volt d-c inverter. It is possible, by changing the bias resistor and size of the wire in the windings, to operate this circuit up to 50 amperes with a 2-volt d-c input. With the parts values shown in the circuit, type 2N1518 transistors are suggested.

In operation assume that Q_1 begins to conduct, which increases the voltage drop across the primary of the transformer. This increasing voltage is coupled to the feedback winding, and a negative voltage is applied to the base of Q_2, causing it to increase its conduction. Q_1 causes the transformer to go into saturation, and the voltage across the primary collapses, causing a change in polarity which is coupled to the feedback winding. The feedback winding applies a reverse bias to Q_1 and a forward bias to Q_2 so that there now is a reverse operation, with increasing conduction through Q_2.

Bias current is applied across R_1. This circuit operates best with a d-c input of 3 volts or less.

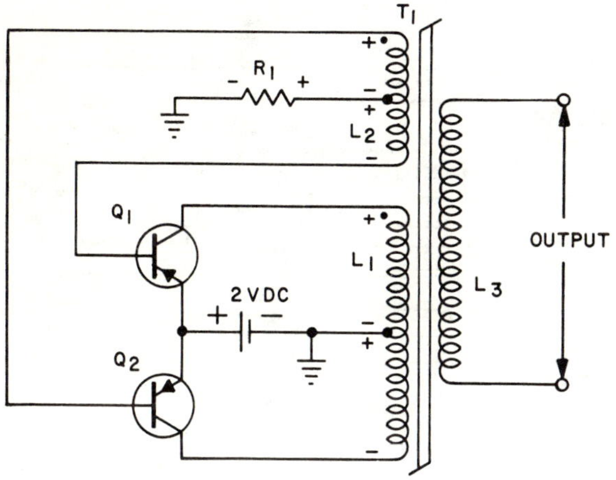

Q_1, Q_2—Delco 2N1518

R_1—7.5 ohm in parallel with 1 ohm = .88 ohm, 10 watt

T_1—L_1 (Primary) 30 turns No. 10, tap at 15 T

L_2 (Feedback) 38 turns No. 19, tap at 19 T

L_3 (Secondary) 1170 turns No. 26

Core: Magnetics Inc. 51001-2A, Toroid. Primary and feedback windings are bifilar.

Courtesy Delco Radio Div., General Motors Corp.

2-volt dc-to-ac inverter.

110-VOLT INVERTER

This circuit shows a dc-to-ac inverter that produces 110 watts of 110-volt 60-cycle output from a 12.6-volt input. In this circuit the transistors operate as switches, with one being off while the other is on.

Assume for the sake of operation that Q_1 is on and Q_2 is off, thus connecting the 12.6-volt battery supply across the top half of the center-tap primary winding. When transistor Q_1 begins to conduct, there is a current flow through the transformer. A saturation of the core causes a collapse of the magnetic field and a switching of the transistor positions.

In this circuit resistors R_1 through R_4 are used to start the oscillations, while resistor R_5, capacitor C_1 and diodes CR_1 and CR_2 are used to provide a de-spiking network to protect the transistors. The suggested heat sink is given in the parts list for this unit.

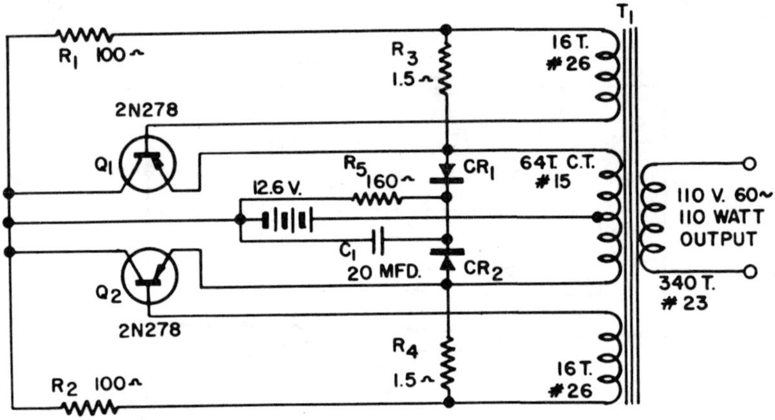

R_1, R_2—100 ohm, 5 watt wirewound
R_3, R_4—1.5 ohm, 5 watt, wirewound
R_5—150 ohm, 5 watt, wirewound
CR_1, CR_2—Power diode, Sarkes Tarzian M 500 or equivalent
C_1—20 mfd, 50V electrolytic
Q_1, Q_2—2N278 Delco power transistor

T_1—transformer: core—1 5/16" stack of 125 E.I. 0.014" silicon iron
Heat sink—aluminum, 80 sq. in., finned. The thermal resistance of this sink should be better than 2.1°C/watt. This will permit operation in ambient temperatures up to 71°C.

Courtesy Delco Radio Div., General Motors Corp.

110-volt inverter.

POWER INVERTER

This circuit shows a power inverter operating from a 12-volt source and using two 2N671 transistors to produce a power output of ½ watt at a frequency of 1380 cycles per second. As shown, the power output is obtained across a 600-ohm load.

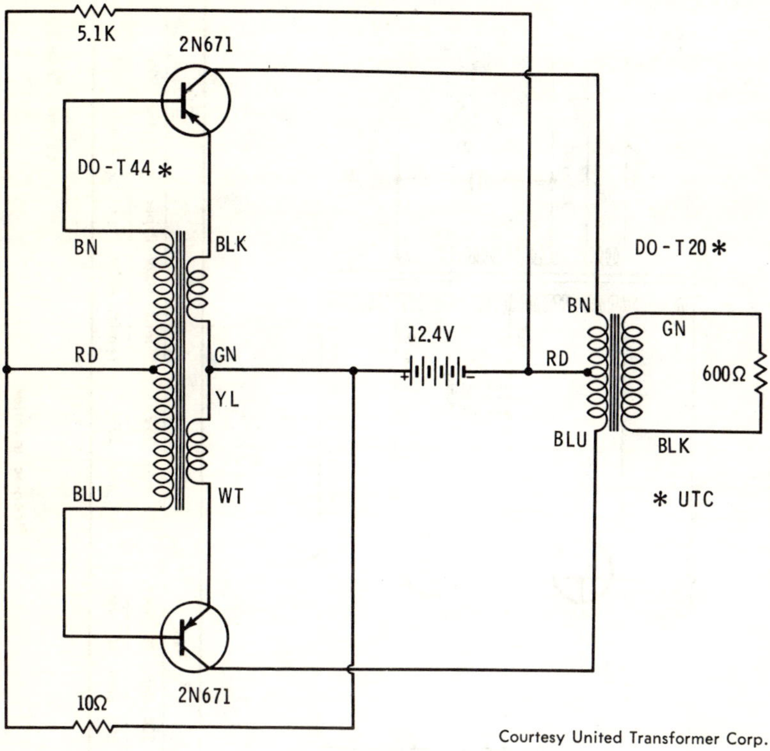

Courtesy United Transformer Corp.

Power inverter.

DC-TO-AC INVERTER

A dc-to-ac inverter using two different transformer types (Triad TY75A or TY76A) is shown. Suggested rectifiers are 1N1763, 1N1764, 1N3196, and 1N3256. Suggested transistors are 2N173, 2N174, 2N443, and 2N1099.

As shown in the circuit, with either transformer the input is 12.6 volts dc. For the TY75A, the output is 115 watts at 60 cycles with voltages of 110, 115, or 125 volts ac. For the TY76A, the output is 60 watts at 60 cycles with outputs of 110, 115, or 125 volts ac.

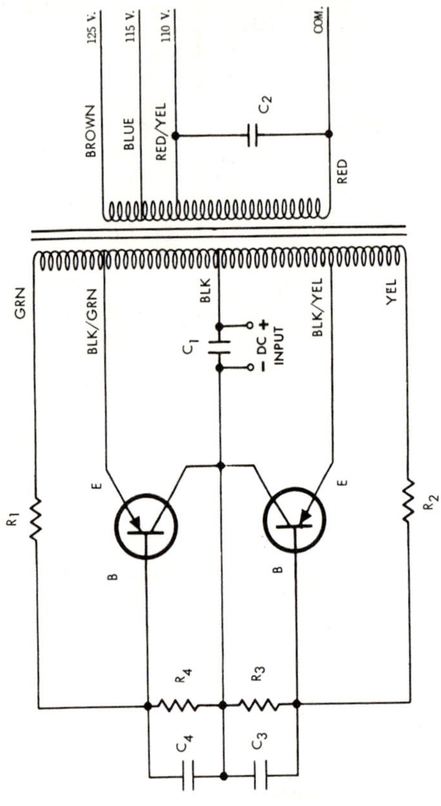

	R_1, R_2	R_3, R_4	C_1	C_2	C_3	C_4
TY-75A	5 ohm, 2 watt	200 ohm, 10 watt	250 mfd, 25V	1 mfd, 400V Paper	2 mfd, 50V	2 mfd, 50V
TY-76A	7 ohm, 2 watt	100 ohm, 10 watt	250 mfd, 25V	2 mfd, 400V Paper	4 mfd, 50V	4 mfd, 50V

Courtesy Triad Transformer Corp.

Dc-to-ac inverter.

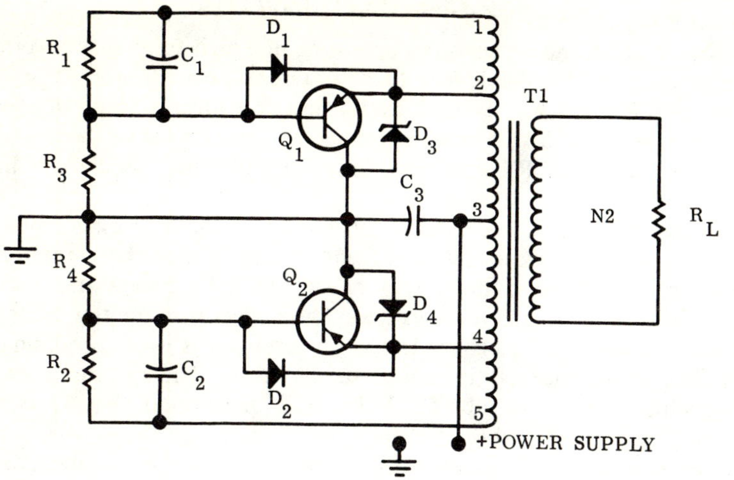

(A) Circuit diagram.

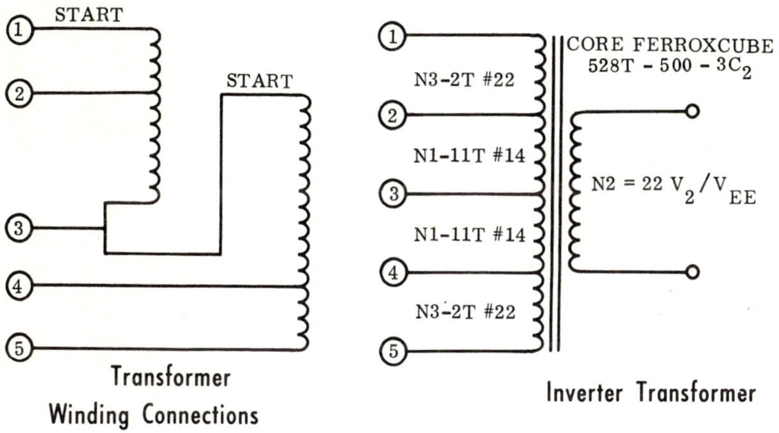

(B) Transformer specifications.

R_1, R_2—27 ohm, 1 watt
R_3, R_4—2.2K, 2 watt
C_1, C_2—0.5 mfd, 10V
C_3—250 mfd, 50V
D_1, D_2—1N4001
D_3, D_4—1N3039 B
Q_1, Q_2—2N2833
T_1—See Fig. B
Power Supply—28V dc, 5 amperes for 100 watts out

Courtesy Motorola Semiconductor Products

High-frequency inverter.

HIGH-FREQUENCY INVERTER

A pair of 2N2833 transistors operating in a high-frequency inverter circuit convert 28 volts dc to a 15-kc square wave as shown in Fig. A. Resistors R_1 and R_3, as well as R_2 and R_4, form voltage dividers which provide 0.35 volt of base voltage necessary to start the circuit. Resistors R_1 and R_2 limit the base current once oscillation has started.

Diodes D_1 and D_2 are used in this circuit as emitter-base clamping diodes which are necessary to prevent reverse bias of the emitter-to-base junction beyond the rating of the 2N2833 transistors. Zener diodes D_3 and D_4 are used to keep the collector-to-emitter voltage of the transistors at a proper level. Capacitor C_3 is used to eliminate the lead inductance of the power-supply leads. The inverter uses a core which is a Ferroxcube 528T-500-3C2; the primary has 11 turns of the No. 14 wire and the feedback winding has 2 turns.

Fig. B shows the inverter transformer and its winding connections.

Section 5

MISCELLANEOUS CIRCUITS

VOLTAGE DOUBLERS

Typical of silicon rectifiers that can be used for voltage-doubler service are the 1N1763 and 1N1764; their operation characteristics are given in these circuits.

Fig. A shows the type 1N1763 in a half-wave voltage-doubler circuit. As is characteristic with voltage doublers, the output voltage decreases as the current drain increases. The unloaded supply in this case has an output voltage of 300 volts dc; at 400 ma drain, the output voltage has dropped to 250 volts using the smaller capacitor and to 270 volts using the larger capacitor.

Fig. B shows the type 1N1764 in a transformer-type half-wave voltage-doubler circuit. Here, with no load, the output voltage is about 425 volts; at 400 ma the voltage has dropped to 320 volts using the smaller capacitor and to 350 volts using the larger capacitor.

Fig. C shows the 1N1763 in a full-wave voltage-doubler circuit fed directly from the line. With no load the output is 330 volts; at 400 ma the output voltage is 240 volts to 260 volts, depending on the size of the output capacitor used.

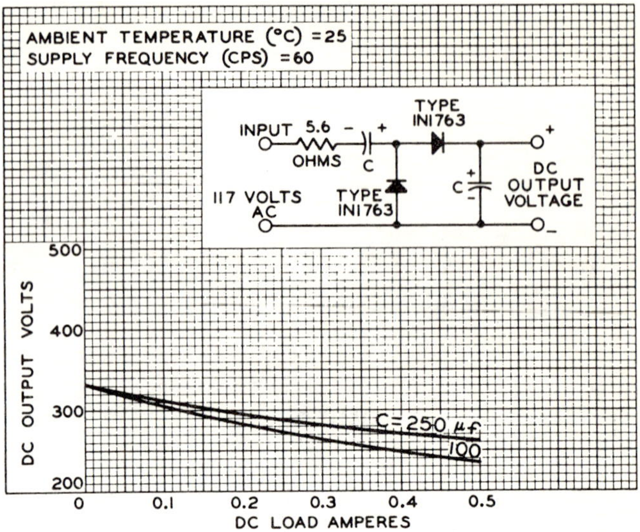

(A) Half-wave voltage doubler.

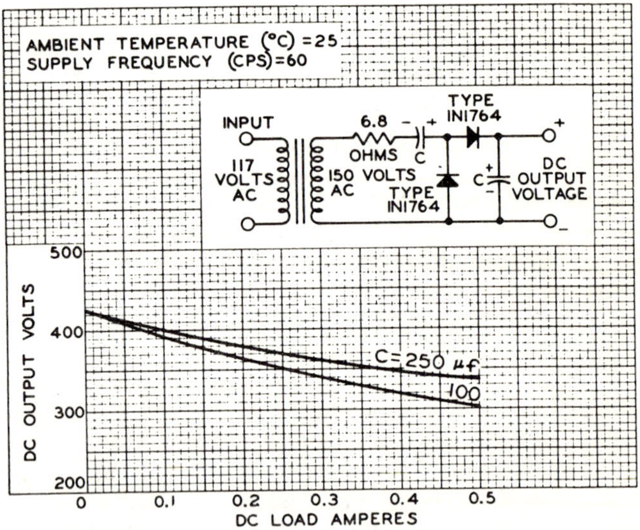

(B) Transformer-input half-wave voltage doubler.

Four typical

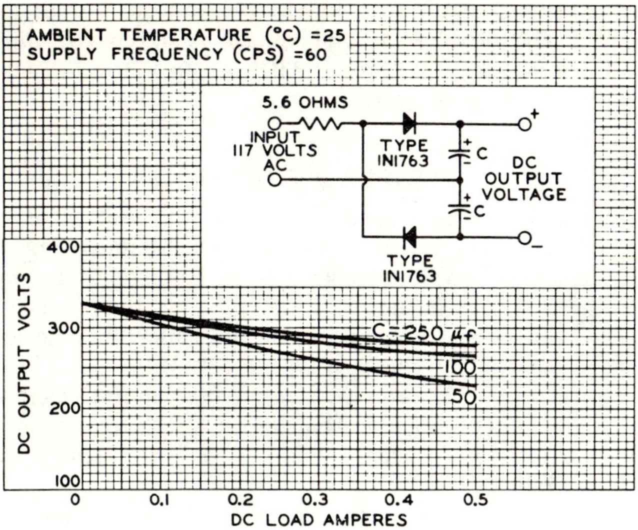

(C) Full-wave voltage doubler.

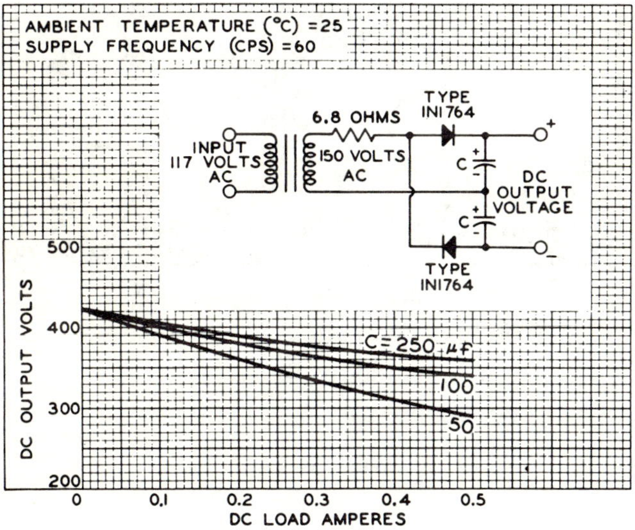

(D) Transformer-input full-wave voltage doubler.

Courtesy Radio Corporation of America

voltage doublers.

Fig. D shows a 1N1764 in a full-wave voltage doubler using a transformer. The unloaded output voltage is 420 volts, while the loaded output voltage at 400 ma is 310 volts to 370 volts, again depending on the size of the capacitor used.

VOLTAGE MULTIPLIER

This circuit is the equivalent of a vibrator circuit as used with an automobile battery to obtain d-c voltages for operating mobile radio equipment or audio amplifiers in an automobile. In this circuit the source voltage is shown as 1.5 volts, but the output voltage and current can be changed by varying the battery voltage and the transformer turns ratio. This is a typical circuit.

Two power transistors are used as an oscillator which operates as soon as the switch is closed. The step-up transformer provides an a-c voltage across the secondary; this voltage is rectified to produce a 50-volt d-c output.

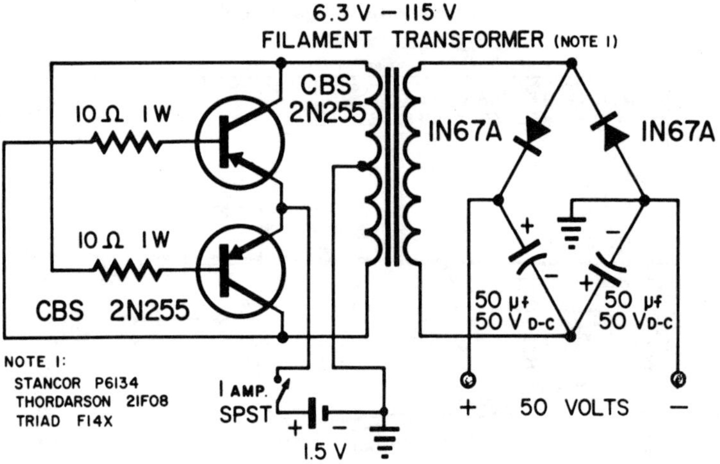

Courtesy CBS Electronics Div., CBS Inc.

Voltage multiplier.

BATTERY CHARGER

A regulated 12-volt battery charger designed to be used with a 12-volt automobile battery is shown. The circuit is set for the required

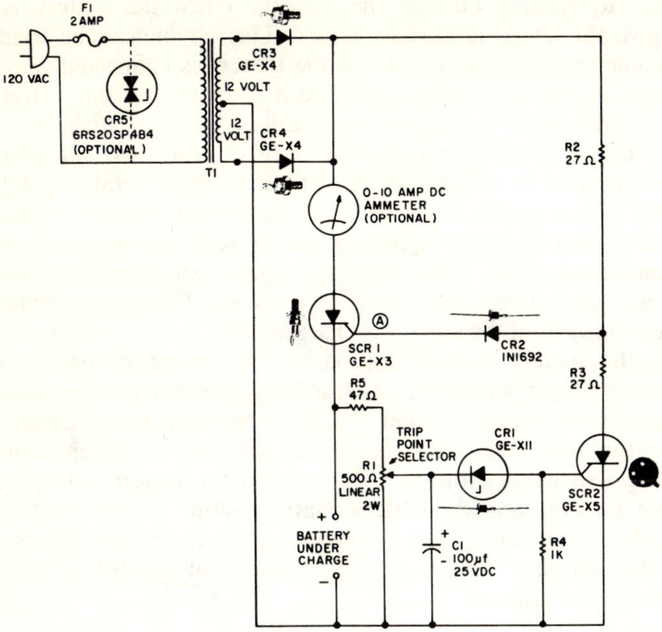

C1—100 μf, 25 volt capacitor
CR1—GE-X11 Zener diode
CR2—G-E Type 1N1692 rectifier diode
CR3, CR4—GE-X4 rectifier diode
CR5—G-E Type 6RS20SP4B4 Thyrector diode (optional transient voltage suppressor)
F1—2 amp fuse
R1—500 ohm, 2 watt linear potentiometer
R2, R3—27 ohm, 3 watt resistor
R4—1000 ohm, ½ watt resistor
R5—47 ohm, 1 watt resistor
SCR1—GE-X3 Silicon Controlled Rectifier
SCR2—GE-X5 Silicon Controlled Rectifier
T1—Transformer: primary, 120 volts AC; secondary, 24 volts AC center-tapped (UTC-FT10, Triad F41X, or equivalent)

Courtesy General Electric Co.

Battery charger.

current to be sent through the battery. When the battery is fully charged, the charging current ceases. This circuit can be used with emergency power-supply sinks. If the battery is connected to the circuit and it becomes discharged while it is connected, the charger will switch back into the circuit automatically.

As shown, there is a full-wave rectifier using two silicon diodes across the 24-volt a-c secondary, with controlled rectifier SCR1 connected in series with the battery. If the battery voltage is less than 12 volts, there is a gate signal applied to SCR1 by means of resistor R2 and diode CR2. With this gate signal applied to the controlled rectifier, the rectifier conducts on each cycle of the supply voltage, and current flows until the battery is charged.

As the battery voltage approaches its normally charged value, capacitor C1 is sufficiently charged to turn on the second semiconductor-controlled rectifier (SCR2) through zener diode CR1. When this point is reached, the voltage at point A, which is the gate of SCR1, suddenly drops to a value below the battery voltage. SCR1 cannot turn on, and the battery charging stops at this point.

With a fully charged battery connected to the circuit, R1 is set so that the charging just ceases. This resistor, or control, is known as the trip-point selector.

SELENIUM-RECTIFIER CIRCUITS

Six basic rectifier circuits and their characteristics are shown. Fig. A is a single-phase half-wave rectifier used for low-power applications. There is a high ripple frequency which is the same frequency as the supply. The efficiency is relatively low.

Fig. B shows the single-phase full-wave bridge for a relatively high power with an efficiency of about 75%. The ripple frequency is twice that of the supply frequency, and this circuit makes greater utilization of the voltage available across the secondary.

Fig. C shows a single-phase full-wave center-tap circuit which requires a transformer with a center tap and a terminal voltage of 2.7 times the load voltage.

Fig. D shows a three-phase half-wave rectifier that can be used where low ripple is of primary importance. The ripple factor is about 20% and the ripple frequency is three times the fundamental frequency. The efficiency of this circuit is higher than the single-phase half-wave circuit.

Fig. E shows a three-phase full-wave bridge circuit that can be used where heavy d-c current is required. The ripple is very small,

SINGLE PHASE

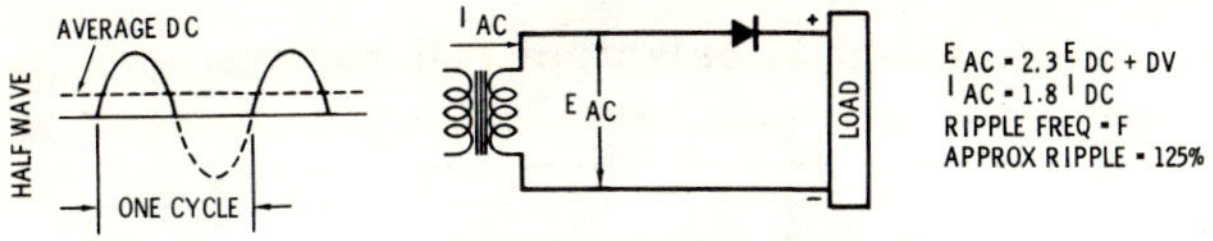

(A) Single-phase half-wave circuit.

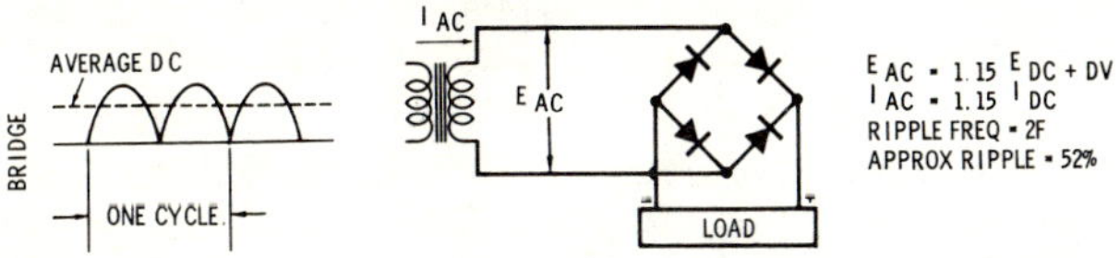

(B) Single-phase full-wave bridge.

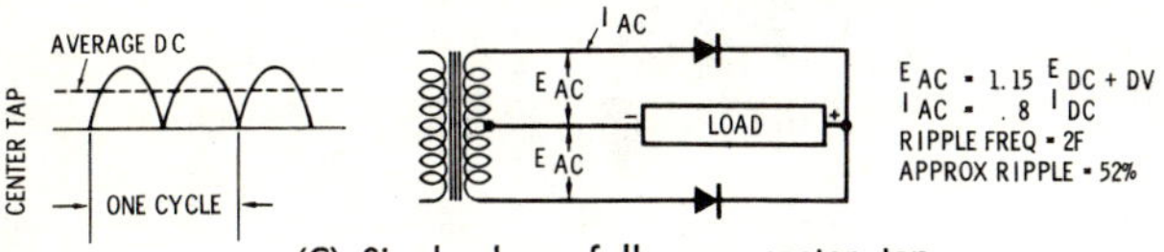

(C) Single-phase full-wave center tap.

THREE PHASE

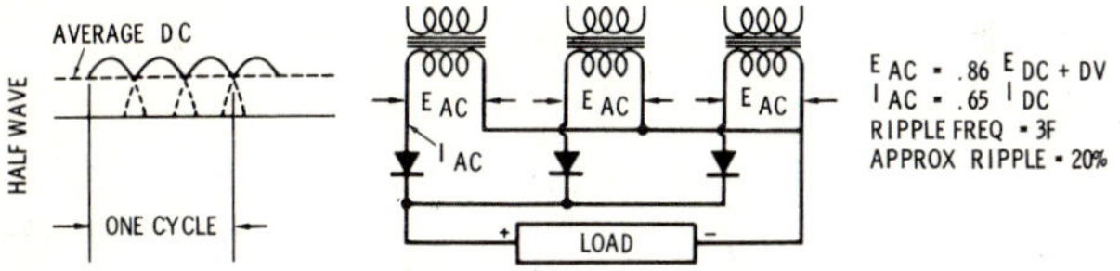

(D) Three-phase half-wave rectifier.

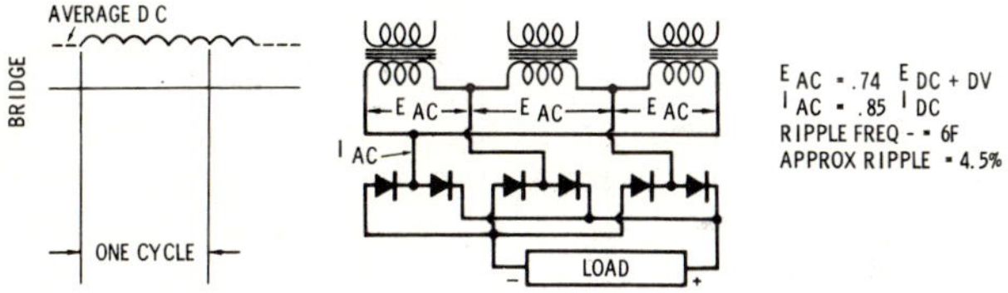

(E) Three-phase full-wave bridge.

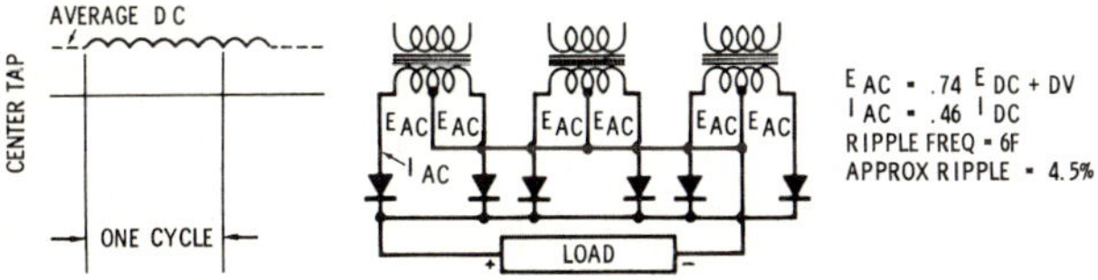

(F) Three-phase full-wave center tap.

E_{AC} = APPROX AC INPUT VOLTAGE (RMS)
E_{DC} = AVERAGE DC OUTPUT VOLTAGE
DV = VOLTAGE DROP IN RECTIFIER (RMS)
I_{DC} = AVERAGE DIRECT CURRENT OUTPUT
I_{AC} = RMS ALTERNATING CURRENT
F = INPUT FREQUENCY
% RIPPLE = $\frac{\text{RMS AC RIPPLE COMPONENT}}{\text{AVERAGE DC VOLTAGE}}$

Courtesy Sarkes Tarzian, Inc.

Selenium rectifier circuits.

Table 1. Selenium cell ratings.

CELL NO.	CELL SIZE	SPAC-ING *	MAX. CELLS PER STACK	CONTINUOUS D. C. AMPERES @ 45°C					
				SINGLE PHASE		THREE PHASE			D. C.
				Half Wave	Bridge and C. T.	Half Wave	Bridge	C. T.	Block-ing
0	.282" D.	...	175	.005	.010	.013	.015	.018	.0075
1	.465" SQ.	...	175	.025	.050	.066	.075	.090	.037
2	1" SQ.	N	30	.075	.150	.2	.225	.27	.12
		N	16	.1	2	.265	.3	.36	.16
		N	8	.112	.225	.3	.34	.4	.18
		W	20	.112	.225	.3	.34	.4	.18
3	1¼" SQ.	N	30	.15	.3	.4	.45	.55	.23
		N	16	.2	.4	.53	.6	.73	.31
		N	8	.23	.45	.6	.675	.82	.35
		W	20	.23	.45	.6	.675	.82	.35
4	1.6" SQ.	N	32	.3	.6	.8	.9	1.1	.45
		N	16	.375	.75	1.0	1.125	1.37	.5
		N	8	.45	.9	1.2	1.35	1.65	.67
		W	24	.45	.9	1.2	1.35	1.65	.67
4A	1.4" SQ.	N	32	.225	.45	.6	.675	.8	.34
		N	16	.275	.55	.75	.85	1.0	.425
		N	8	.35	.7	.9	1.15	1.2	.5
		W	24	.35	.7	.9	1.15	1.2	.5
5	2" SQ.	N	32	.5	1	1.33	1.5	1.8	.75
		N	16	.6	1.2	1.6	1.8	2.18	.86
		N	8	.75	1.5	2	2.25	2.7	1.1
		W	24	.75	1.5	2	2.25	2.7	1.1
5A	1.8" SQ.	N	32	.4	.8	1	1.2	1.4	.6
		N	16	.48	.96	1.2	1.45	1.7	.7
		N	8	.6	1.2	1.5	1.8	2.1	.9
		W	24	.6	1.2	1.5	1.8	2.1	.9
6	3" SQ.	N	32	1	2	2.7	3	3.6	1.5
		N	16	1.165	2.33	3.14	3.5	4.2	1.75
		N	8	1.25	2.5	3.38	3.75	4.5	1.88
		W	20	1.5	3	4	4.5	5.4	2.25
7	4" SQ.	N	32	2	4	5.3	6	7.2	3.2
		N	16	2.25	4.5	5.9	6.75	8.1	3.6
		N	8	2.5	5	6.6	7.5	9	4
		W	20	3	6	8	9	10.8	4.8
8	5" SQ.	N	40	3.5	7	9.3	10.5	12.6	5.6
		N	16	3.9	7.9	10.4	11.8	14.1	6.3
		N	8	4.2	8.3	11.1	12.5	15	6.7
		W	24	5.3	10.5	14	15.8	18.5	8.4
9	4¼"x6"	N	40	3.5	7	9.3	10.5	12.6	5.6
		N	16	3.9	7.9	10.4	11.8	14.1	6.3
		N	8	4.2	8.3	11.1	12.5	15	6.7
		W	24	5.3	10.5	14	15.8	18.5	8.4
10	5" x 6"	N	40	4	8	10.6	12	14.4	6.4
		N	16	4.5	9	1.9	13.5	16.2	7.2
		N	8	4.75	9.5	12.5	14.25	17.1	7.6
		W	24	6	12	16	18	21.6	9.6
11	4½" x 6⅝"	N	40	4	8	10.6	12	14.4	6.4
		N	16	4.5	9	11.9	13.5	16.2	7.2
		N	8	4.75	9.5	12.5	14.25	17.1	7.6
		W	24	6	12	16	18	21.6	9.6
12	4¼" x 12"	N	40	7.5	15	20	22.5	27	12
		N	16	8.5	16.9	22.4	25.3	30.4	13.5
		N	8	8.9	17.8	23.9	26.9	32	14.3
		W	24	11.2	22.5	30	33.8	40.5	18
14	6¼" x 7¼"	N	40	6	12	15	17	20	9
		N	16	6.5	13	16.9	18	21	10.5
		N	8	7	14	17.8	19	22	12
		W	24	9	18	22.5	24	30	14
*N—Normal Spacing W—Wide Spacing				Maximum RMS Volts per Cell 33V					Max. Blocking Volts 27.5 Vdc

Single phase current ratings are for resistive or inductive loads only. For battery, motor, or capacitive loads the ratings are 80% of values shown above. All ratings are based on normal convection cooling.
For fan cooling (approximate air velocity of 500 feet per minute) multiply normal spaced ratings by 2.5.

Courtesy Sarkes Tarzian, Inc.

about 4%, and additional filtering is often not required. This circuit has a d-c voltage output that is about 20% higher than the applied single-phase voltage.

Fig. F shows a three-phase full-wave center-tap circuit to be used where extremely heavy d-c currents are required. A transformer with a six-phase star secondary is needed, with each phase winding of the secondary provided with a center tap.

Suggested selenium cells for use with these circuits are shown in Table 1, with the characteristics given for each cell for each circuit type.

SWITCHING REGULATOR

A switching type of voltage regulator is shown in block diagram form in Fig. A and in schematic form in Fig. B. The function of a switching regulator is to reduce the power loss in the series transistors when there is a large variation of the input or output voltage. A switching regulator operates on the principle that the output voltage is proportional to the input voltage according to the duty factor of the switch voltage.

There is a series control which is effectively a variable resistance feeding an output filter. The voltage output is compared to a reference, and then the error voltage or difference is used to adjust the duty cycle control. This feeds a driver that in turn adjusts the series control resistance.

The amplifier stage is made up of a differential amplifier and a stage of current gain; Q_1 and Q_2 make up the differential amplifier, while Q_3 is the stage for additional gain. The purpose of the differential amplifier is to sense the output voltage and compare this to a reference so that the difference voltage can be amplified. The amplified error voltage is applied to the base of transistor Q_3 which controls the charging rate on capacitor C_3.

A zener diode D_1 provides a stable voltage source used for a reference, while zener diode D_2 provides an offset voltage for the reference and differential amplifier. This allows the output voltage to be variable from 0 to 20 volts. Without this offset voltage, the minimum output voltage would be the reference voltage.

Transistors Q_4 through Q_8 provide the duty-cycle control circuit. Unijunction transistors Q_5 and Q_8 provide a relaxation oscillator for duty-cycle control and for frequency control. The oscillators produce pulses to trigger a bistable multivibrator circuit. Q_8 controls the operating frequency of the regulator at 5 kc. Transistor Q_5 sets the

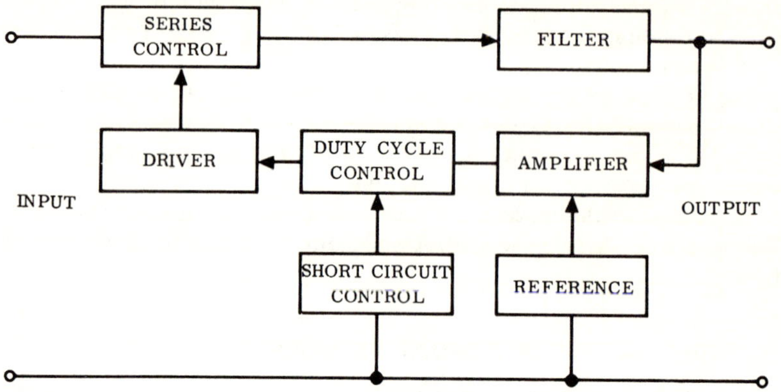

(A) Block diagram.

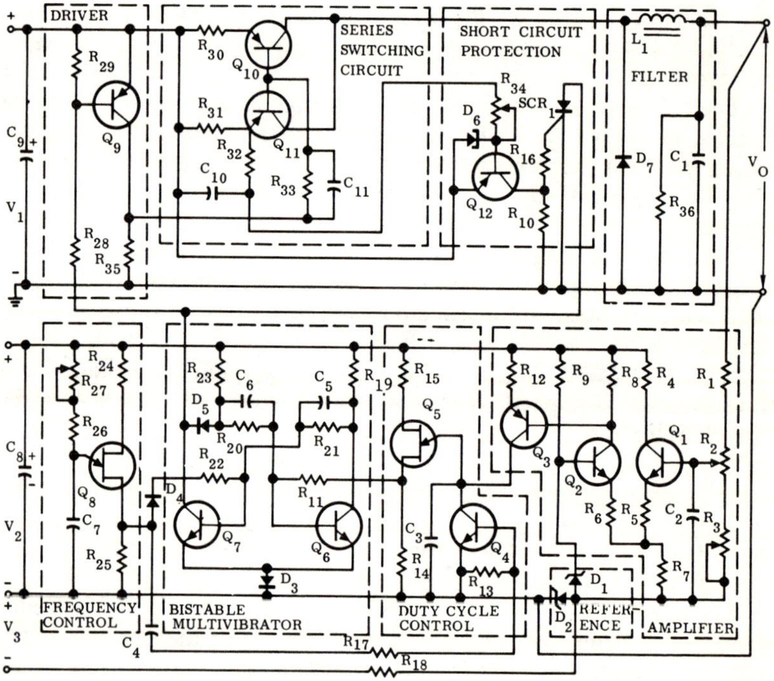

(B) Schematic diagram.

Switching

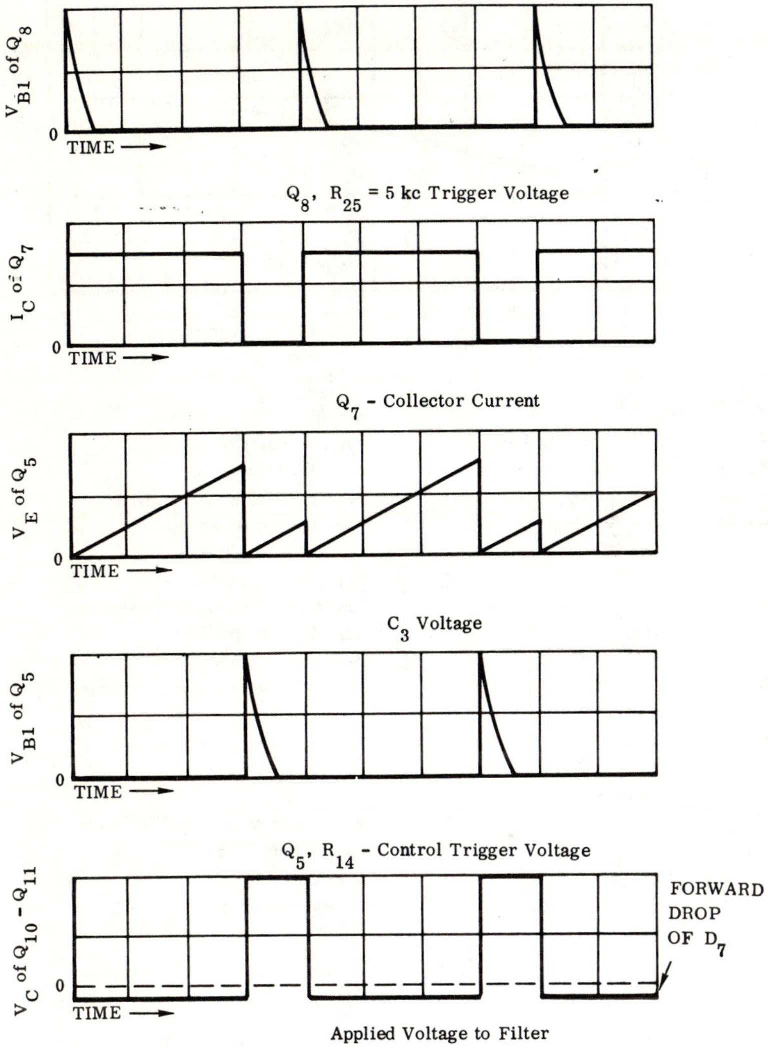

(C) Circuit wave shapes.

Courtesy Motorola Semiconductor Products

regulator.

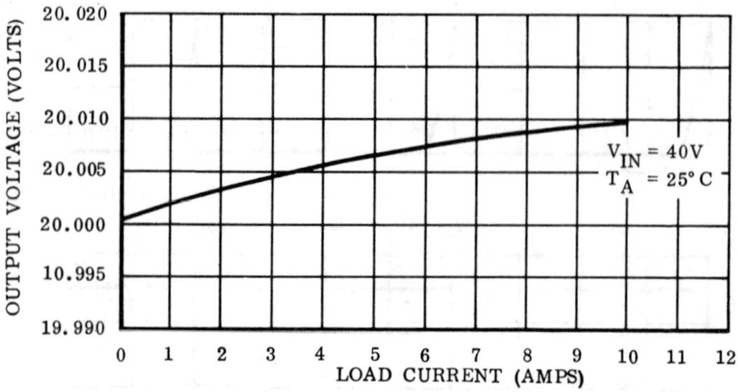

(D) Load regulation.

R_1—120 ohm, ±1%
R_2—500 ohm, 10T potentiometer
R_3—5 ohm potentiometer
R_4—2.2K, ±1%
R_5, R_6—100 ohm, ±1%
R_7—12K, ±1%
R_8—10K, ±1%
R_9—2.7K, ±1%
R_{10}, R_{13}, R_{16}, R_{18}—4.7K, ±5%
R_{11}, R_{14}, R_{25}—100 ohm, ±5%
R_{12}—150 ohm, ±1%
R_{15}, R_{24}—330 ohm ±5%
R_{17}—5.6K, ±5%
R_{18}, R_{21}—1.8K, ±5%
R_{19}, R_{20}, R_{23}—1K, ±5%
R_{22}—470 ohm, ±5%
R_{26}—36K, ±5%
R_{27}—10K potentiometer
R_{28}—330 ohm, 5 watt, ±5%
R_{29}—20 ohm, ±5%
R_{30}, R_{31}—0.1 ohm, 5 watt, wirewound
R_{32}—47 ohm, ±5%
R_{33}—2 ohm, 5 watt, wirewound
R_{34}—100 ohm, 10T potentiometer
R_{35}—40 ohm, 50 watt, wirewound
R_{36}—100 ohm, 5 watt, wirewound

All resistors are ½ watt unless otherwise specified.

C_1—12,500 mfd, 75V
C_2—0.001 mfd
C_3—0.003 mfd
C_4—0.05 mfd
C_5—2200 pf
C_6—.01 mfd
C_7—0.006 mfd
C_8—50 mfd, 25V
C_9—2000 mfd, 75V
C_{10}—0.1 mfd
C_{11}—50 mfd, 15V

All capacitors are ±10%.

D_1, D_2—1N823A
D_3, D_4, D_5—1N4001
D_6—1N3128
D_7—1N3889
Q_1, Q_2, Q_4, Q_6, Q_7—2N697
Q_3—2N2800
Q_5, Q_8—MU970
Q_9, Q_{10}, Q_{11}—2N2832
Q_{12}—2N2043
SCR-1—MCR 914-1
L_1—8 mh at 10A dc. Triad C-49U
V_1—45V ±5V at 10 amps peak with a 50% duty cycle
V_2—20V ±5% at 75 ma
V_3—20V ±5% at 25 ma

Courtesy Motorola Semiconductor Products

Switching regulator. (Cont)

"on-off" duty cycle of the 5 kc rate by generating pulses at time intervals which depend on the rate of charge of capacitor C_3.

Transistor Q_4 discharges capacitor C_3 at the beginning of each trigger pulse, thus resetting the charge on this capacitor from zero at the beginning of each cycle. Transistors Q_6 and Q_7 constitute a bistable multivibrator; pulses from the unijunction circuits trigger this multivibrator to produce variable duty-cycle pulses. Diode D_3 provides bias for the transistors.

Power transistor Q_9 is used as the driver circuit which is driven into saturation by the collector current of Q_7. Saturating Q_9 causes the series switching transistors to be biased "off" due to the charge on capacitor C_{11}.

Transistors Q_{11} and Q_{10} are the series switching resistors operating in parallel; they are biased "on" through resistor R_{35} and turned "off" by the driver transistor.

Circuit waveshapes are shown in Fig. C, and the load regulation for this circuit is shown in Fig. D. This load regulation from no-load to full-load is 0.05%.

ELECTRONIC RIPPLE FILTER

It is possible to use transistor circuits to reduce the ripple in a power-supply filter, but these circuits are not designed to take the place of ordinary inductance and capacitance filters. They are designed to supplement the ordinary filter to produce very low values of ripple. For a circuit such as this to operate efficiently, it is necessary to reduce the ripple to an approximate value of 3 volts peak-to-peak or less, ahead of the electronic filter.

Consider Fig. A as the basic electronic ripple-filter circuit which can be redrawn as an emitter follower as shown in Fig. B. The emitter follower, or common-collector circuit, is degenerative, and any increase in voltage at the input will cause a decrease in voltage across the load so that the ripple is effectively reduced by the degeneration process. Another name for this circuit is the capacitance multiplier. Suppose that in Fig. B a ripple voltage appears which would add to the 31 volts shown. This ripple voltage would have a tendency to increase the voltage across the load resistor, but note that the polarity of the increase will actually cause the base of the transistor to become less negative with respect to the emitter. This has a tendency to turn off the transistor, and the ripple voltage subtracts from the d-c power supply so that, in effect, any increase in voltage caused by the ripple will be effectively counteracted.

In Fig. A the capacitor shown in the circuit will try to maintain a nearly pure d-c voltage just as a battery would in the circuit. However, the capacitor has the advantage over the battery in that it can change its d-c voltage level, if required by the input voltage level.

Fig. C shows a circuit that will reduce the ripple by about a factor of 250 to 1 so that if the ripple is reduced to about 250 millivolts (peak to peak) by the inductance and capacitance filters ahead of the circuit, the output from this electronic filter circuit will only contain approximately 1 millivolt of ripple peak-to-peak.

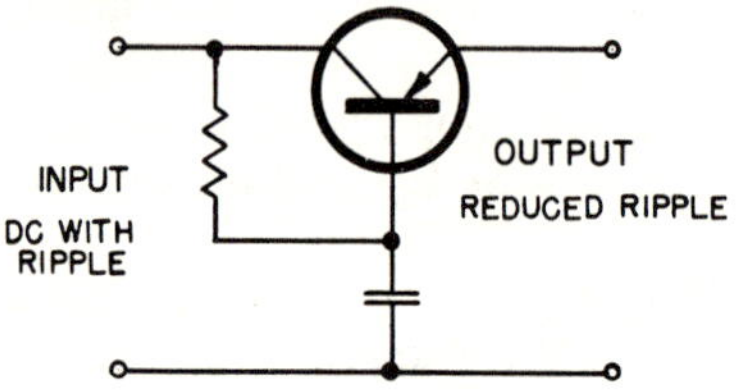

(A) Basic ripple-filter circuit.

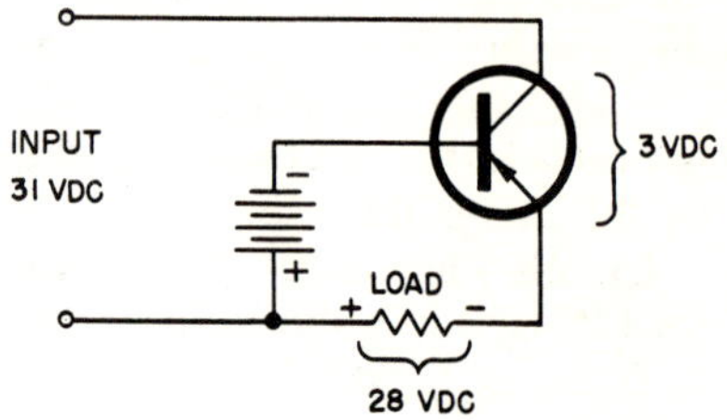

(B) Redrawn as an emitter follower.

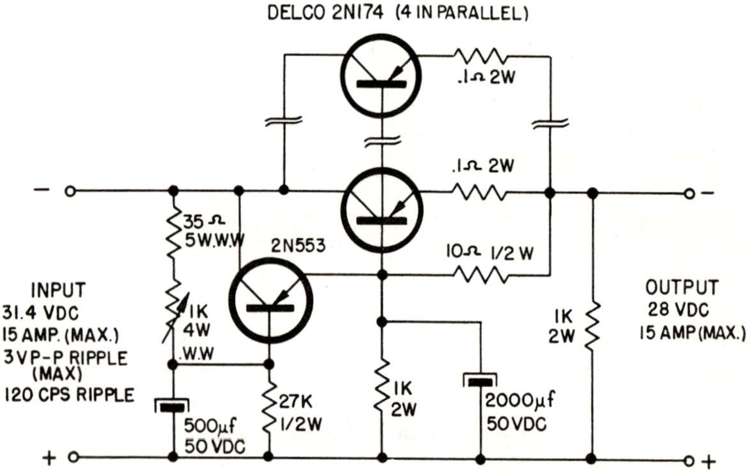

(C) Circuit reduces ripple by 250 to 1.

Courtesy Motorola Semiconductor Products

Electronic ripple filter.

PLUG-IN SILICON RECTIFIERS AS TUBE REPLACEMENTS

It is possible to replace vacuum-tube rectifiers with a direct plug-in replacement made from individual silicon rectifiers. These rectifiers and the tube type for replacement are shown in Table 2. For example, to replace a 5AU4 tube, a 1N2389 is suggested.

Table 2. Silicon-rectifier replacements for tubes.

amps. DC (100°C)	max. per section: peak inverse voltage	max. per section: peak rectifier current	DC output current MA, max.	replacement for type	Tarzian Type	Jedec No.	pins*	dimensions in., max. A	B	C
0.25	4500	2500	250**	6AU4, 6AX4 6BL4, 6W4, 12AX4, 17AX4, 25W4	S-5033	1N1262		3¾	1⅜	4 5/16
.250	19000	2500	250**	8020*	S-5367	—		6¼	2 1/16	6 27/32
0.3	7000	3000	300**	816, 836 or 3B28 and 866 at reduced voltage	S-5343	—		3½	1¼	4 5/32
	10400	3000	300**	866, 866A, 3B28	S-5130	—		4 7/16	1¼	5
0.5	2800	5000	500	5R4	★S-5019	1N1239		3¾	1⅜	4 5/16
	1600	5000	500	6X4	★S-5207	1N2490		1 9/16	13/16	1 13/16
0.6	1600	6000	600	5AU4, 5AW4, 5AZ4 5T4, 5U4, 5V4, 5W4, 5Y3, 5Z4	S-5251	1N2389		1¾	1¼	2¼
0.75	1600	8000	750	80, 82, 83, 83V, 5Z3	S-5011A	1N1150A		2 21/32	1¼	3 7/32
	1600	8000	750	0Z4, 5X4, 5Y4, 6AX5, 6X5	S-5017	1N1237		2 21/32	1¼	3 7/32
	1600	8000	750	5AU4, 5AW4, 5AZ4, 5T4, 5U4, 5V4, 5W4, 5Y3, 5Z4	S-5018	1N1238		2 21/32	1¼	3 7/32
0.5	1600	5000	500	12BW4, 6BW4	★S-5347	—		2	13/16	2¼
1.2	10000	10000	†1250	872A	S-5344	—		7 7/16	1¾	7¾
1.2	10000	10000	†1250	8008	S-5373	—		7 7/16	1¾	8 3/16
1.5	15000	10000	‡1500	575A	S-5449	—		9 11/16	1¾	10

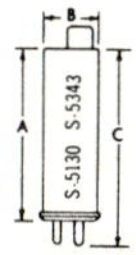

*cathode ●; anode ○. Numbered terminals are wired, others are blank.

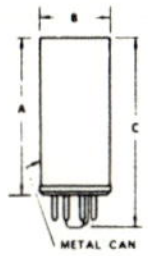

★ Hermetically sealed.
† These Units are Rated at 1,250 MA (Resistive-Inductive-Load; 1000 MA Capacitive Load) with Forced Air Cooling of 1000 LFM. Convection Cooled Rating is 750 MA. (R-I Loads, 600 Capacitive Load).
‡ This Unit is Rated at 1,500 MA R-I Load; 1200 MA Capacitive Load Forced Air Cooling of 1000 LFM. Convection Cooled Rating is 750 MA. (R-I Loads, 600 MA Capacitive Load).
• At Reduced Voltage (8020 has PIV of 40,000 Volts).
•• Rating is for Resistive Load; Derate DC Output Current by 20% for Capacitive Load.

Courtesy Sarkes Tarzian, Inc.

ELECTRONIC RIPPLE FILTER

This circuit shows a simplified electronic ripple filter operating basically on the principle of a common-collector or emitter-follower circuit where the degeneration in the circuit effectively reduces the ripple coming from the power supply.

This electronic ripple filter is connected in a circuit so that the input to the filter is the output of the conventional inductance and capacitance filter. This circuit will reduce the ripple by about 6 to 1, as shown, and requires only three transistors.

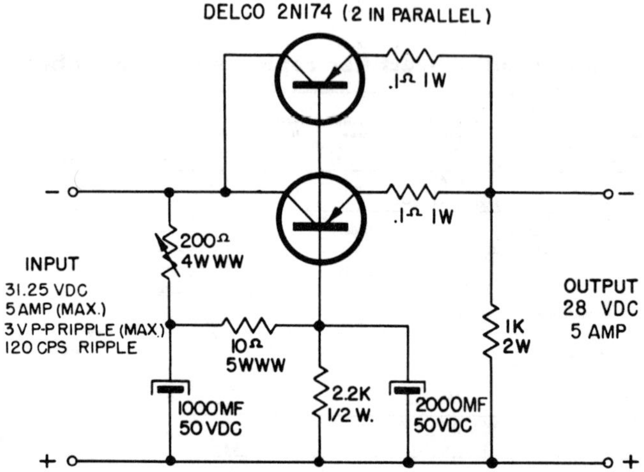

Courtesy Motorola Semiconductor Products

Electronic ripple filter.

STAIRCASE GENERATOR

This circuit is a staircase waveform generator using a unijunction transistor to reduce the number of circuit components. The circuit can generate a staircase waveform at the emitter of unijunction transistor 2N492A, and it can also divide the input frequency by any desired ratio.

As shown in the circuit, a negative input pulse through transistor Q_2 turns this transistor on for a period of 15 microseconds as determined by resistor R_2. A pulse from this conduction provides the drive for diode D_1 and transistor Q_3, which together provide a current source controlled by resistor R_4. Capacitor C, as shown, is charged by this current source. The particular voltage at each step is determined by the value of resistor R_4, and when the voltage across capacitor C reaches the firing potential of the unijunction 2N492A transistor, the unijunction will fire and discharge the capacitor. This discharge produces a positive output pulse across the 47-ohm re-

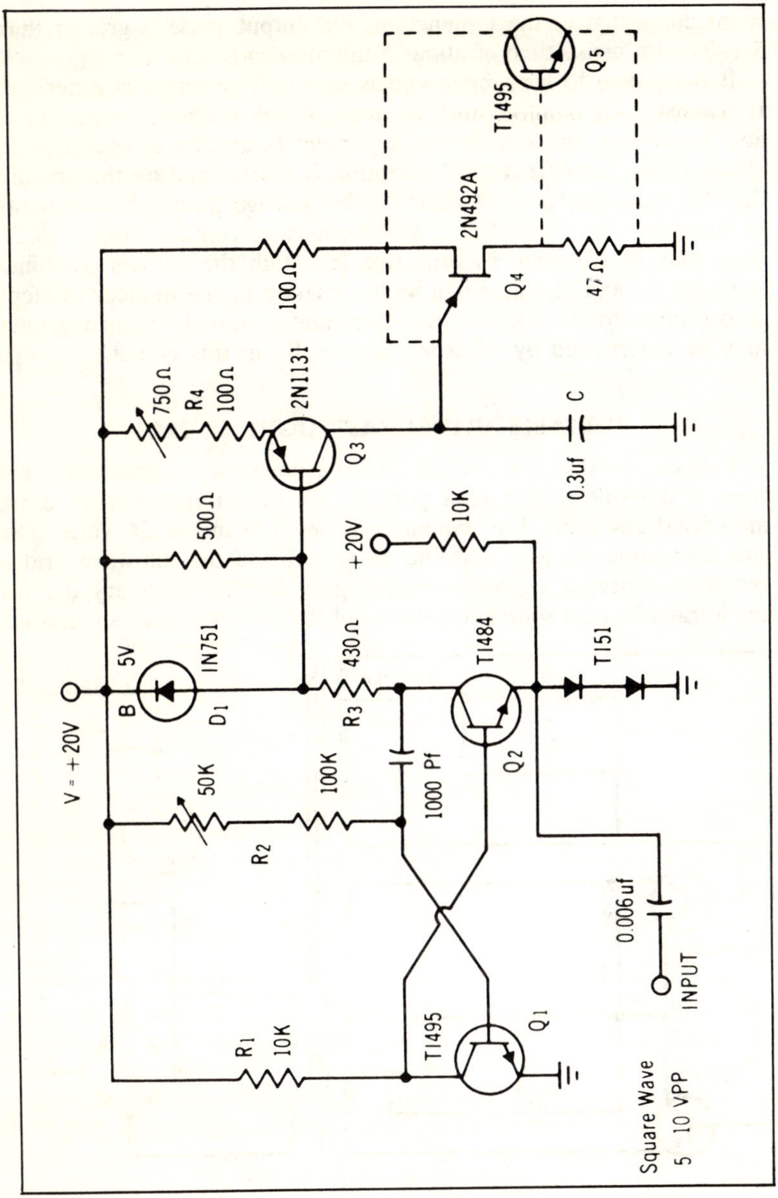

Courtesy Texas Instruments Inc.

Staircase generator.

sistor connected to the unijunction; the output pulse is greater than 5 volts with a rise time of about 5 microseconds.

It is possible to feed input signals up to 10 kc into this generator. If transistor Q_5 is added to the circuit, it will discharge capacitor C at a faster rate and double the maximum frequency of operation to about 10 kc. Low-frequency operation is determined by the amount that the capacitor is discharged by the leakage path, which consists of transistor Q_3, Q_4 and Q_5. Waveforms that contain from 5 to 25 steps may be obtained by adjusting R_4. With the various combinations for R_2 and R_4 there is a wide variation in the number of steps across capacitor C, and the maximum and minimum frequency rates may be determined by adjusting resistor R_2 in this circuit.

PORTABLE ULTRAVIOLET LIGHT SOURCE

This circuit shows a pair of 2N2912 power transistors operating from a 2.5-volt source as a portable ultraviolet light source using an F5000 gas tube. This circuit provides 400 ma at 26 volts. The two transistors act as a dc-to-ac converter, and the full-wave bridge rectifier connected to the secondary provides the necessary d-c for the ultraviolet light source. Operation of the converter is at about 8 kc.

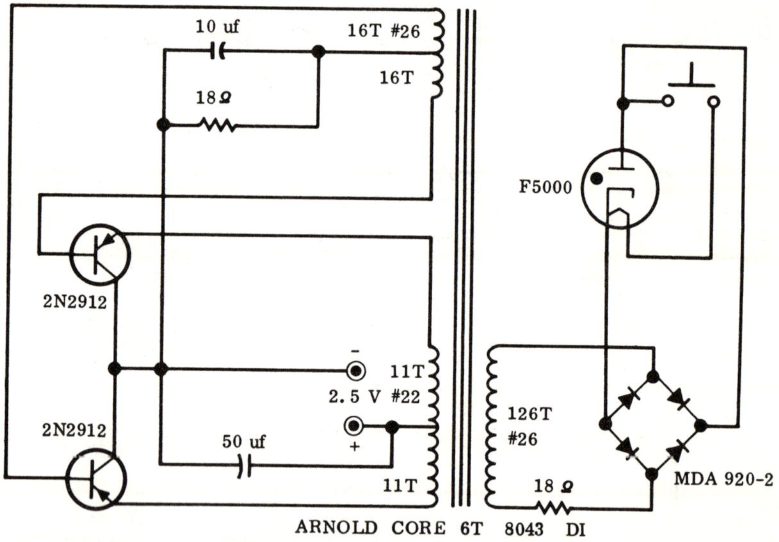

Courtesy Motorola Semiconductor Products

Portable ultraviolet light source.

ELECTRONIC FUSE CIRCUIT

This circuit shows an electronic fuse used to protect power transistors. It is designed to be used from a 7- to 12-volt 5-ampere regulated power supply, and when the load in this case reaches 5.5 amperes, the transistor will open the circuit.

With the application of supply voltage V_1 to the circuit, transistor Q_1 is saturated by the drop across resistor R_3. Load current will cause a voltage drop across resistor R_1 which is in the positive leg of the circuit; when this voltage reaches about 0.8 volt, the diode D_1 and transistor Q_2 will begin to conduct as shown.

Conduction in Q_2 causes the base of Q_1 to be driven more positive, and when the collector of Q_1 is sufficiently negative so that D_2 can conduct, transistor Q_2 will switch on completely and at the same time transistor Q_1 will be turned off.

It is possible to reset the circuit by removing the short or excess load and by opening the reset switch momentarily. It is also possible to reset the circuit by removing the power for a moment.

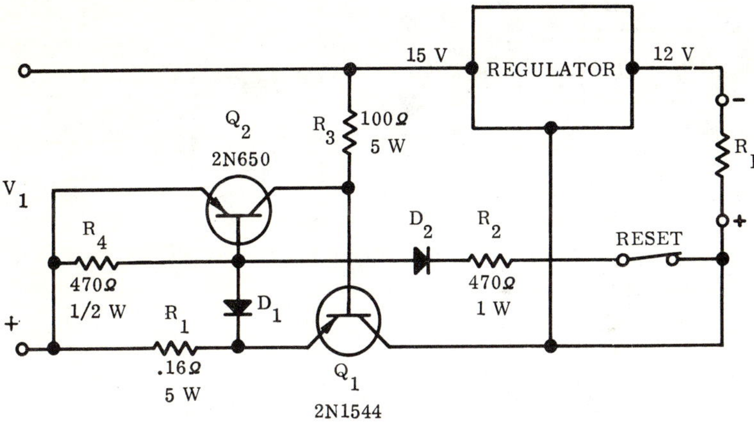

Courtesy Motorola Semiconductor Products

Electronic fuse circuit.

SILICON RECTIFIERS AS TUBE REPLACEMENTS

There are cases where it is desirable or necessary to replace vacuum-tube diode rectifiers with silicon rectifiers. The silicon rectifiers will provide a longer life, they will withstand more physical damage than vacuum tubes, and they take up less room so that a circuit can be miniaturized. It is possible to make a direct substitution for vacuum tube or gas-tube rectifiers with silicon rectifiers by observing certain precautions.

The voltage output of the circuit will increase because the voltage drop of the silicon rectifier is less than that of the vacuum tube or a

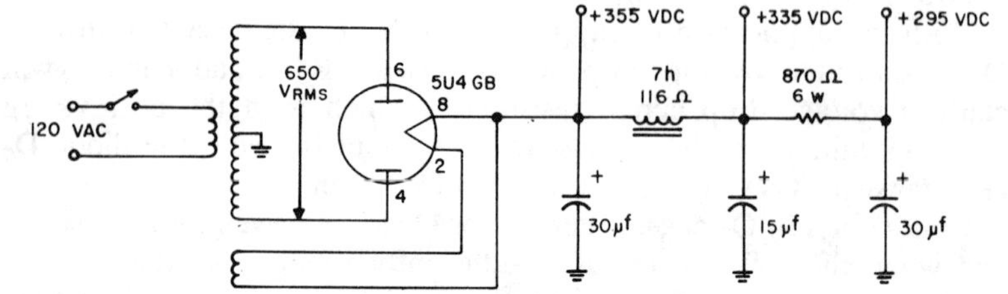

NOTE: PEAK STEADY STATE REVERSE VOLTAGE ACROSS ANODE TO CATHODE OF 5U4GB IS 650$\sqrt{2}$ = 920 VOLTS.

(A) Original circuit.

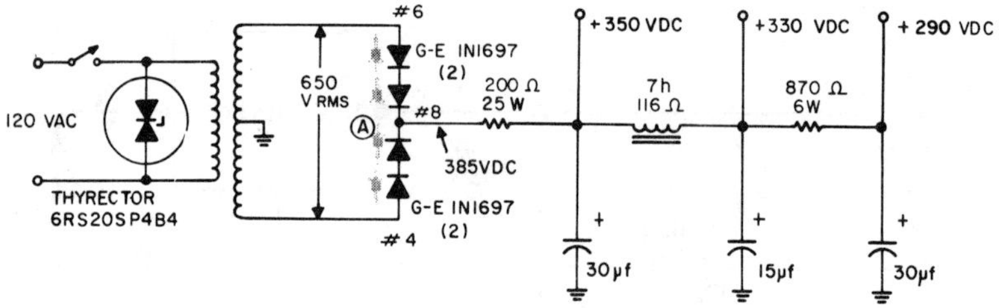

(B) Silicon-rectifier replacement circuit—resistance in load.

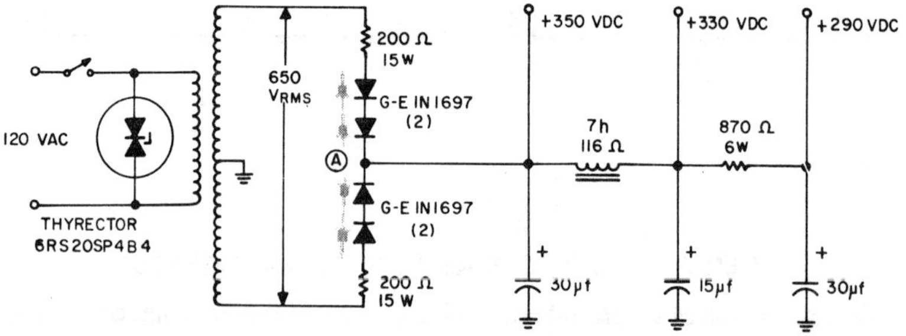

(C) Silicon-rectifier replacement circuit—resistance in each leg.

Courtesy General Electric Co.

Silicon rectifiers as tube replacements.

gas tube. A silicon rectifier will, in general, provide about 10 volts more output than a gas or mercury rectifier, and as much as 50 volts more than the voltage output of a vacuum-tube rectifier.

The increased voltage of a silicon rectifier over a gas or mercury rectifier is normally not a problem in the circuit, but when substituting a silicon rectifier for a vacuum-tube rectifier, the increased voltage output must be compensated for to prevent damage to the circuit which is connected to the power supply.

Fig. A shows a typical vacuum tube, a 5U4GB full-wave rectifier in a power supply; Fig. B of the circuit shows a silicon rectifier replacement with the addition of a 200-ohm resistance to compensate for the increased voltage output; Fig. C shows the additional resistance used in series with the rectifier diodes themselves.

Essentially, the value of this extra load resistance is determined by the characteristics of the tube itself. It is necessary in a particular circuit to find the voltage drop across the tube at the required current output. Using this voltage and this current the tube resistance under these conditions may be calculated, and the load resistance is approximately equal to this value of tube resistance. The power handling capability of the resistor is important, and when the current and the resistance are known, the required wattage rating can be calculated. A safe value of wattage, preferably several watts greater than the required value, should be used.

Another consideration is that there is no warm-up time with a silicon rectifier, and if d-c voltage from the power supply is immediately applied to the vacuum-tube plates in the associated amplifier or other circuit, the vacuum tubes may be damaged. In many types of vacuum-tube equipment, a rectifier tube with an indirectly heated cathode, such as a 6X4 or 35Z6, is used so that the rectifier tube will have approximately the same heating characteristics as the amplifier tubes. If a silicon rectifier is used to replace a vacuum rectifier with an indirectly heated cathode, a time-delay relay or time-delay switch should be used in the circuit; under these circumstances a silicon rectifier can be used in place of an indirectly heated cathode vacuum-tube rectifier. In the circuits of Figs. B and C transient voltage suppressors are suggested to prevent excessive voltage from appearing across the silicon rectifier.

Table 3 shows some suggested silicon rectifier replacements for directly heated vacuum-tube rectifiers. Table 4 shows some silicon rectifier replacements for gas or mercury-vapor tube rectifiers. For example, the tube to be replaced is selected from the first column and its symbol or type is given in the second column. The tube ratings

Table 3. Suggested silicon-rectifier replacements for gas or mercury-tube rectifiers.

TUBE RECTIFIER	TYPE	CATHODE	SYMBOL	TUBE RATING: Peak Reverse Volts	Peak Current Amps	DC Current Amps Per Anode	SILICON RECTIFIER REPLACEMENT (One Diode Per Leg): SYMBOL	Type[1]	Max RMS Voltage of Transformer Sec.[4] (end to end)	Peak Reverse Voltage Rating	Max Transient Voltage Rating	DC Current Max Amps 100° C Ambient	Thyrector For Secondary
OY4 & OY4G	Gas	Cold		300	.500	.075		1N1694 1N1695 1N1696 1N1697 1N560 1N561	180 240 282 325 387 480[2]	300 400 500 600 800 1000[2]	450 600 700 800 960 1200[2]	.250 .250 .250 .250 .250 .250	6RS20SP or 6RS20SC —6B6 —8B8 —9B9 —11B11 —13B13 —16B16
OZ4, OZ4G (OZ4A)	Gas Gas	Cold Cold		880 880	.200 .330	.075 .110							
82 83	Hg Hg	Hot Hot		1550 1550	.6 1.0	.060 .115							
816	Hg	Hot	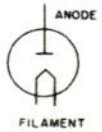	5000	.5	.125							
866/866A	Hg	Hot	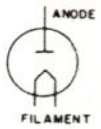	10000 2500	1.0 2.0	.25 .50	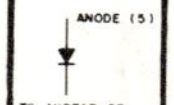	4JA10C 4JA10D 4JA10E 4JA10M	150 210 262 315[2]	300 400 500 600[2]	400 525 650 775[2]	.650[3] .650[3] .650[3] .650[3]	6RS20SP or 6RS20SC —5B5 —7B7 —9B9 —11B11

NOTES:

[1]Any of the silicon rectifiers listed will handle the current. Select on the basis of voltage.

[2]For higher voltages, use rectifiers in series.

[3]Average current reduced to .4 amps for 300 μfd capacitor load and 3.3 ohm surge resistor.

[4]The maximum RMS voltage of the transformer secondary has been calculated from the rating of the Thyrector diode (6RS20SP4E4), if used on the 120 volt AC transformer primary. Transient voltages will be suppressed below 175% of the rated peak ($\sqrt{2}$ RMS) transformer primary and secondary voltages.

[5]A 15 to 20 second time delay of the rectified power output is suggested where one is not already present to protect other vacuum tubes in the circuit from the possible detrimental affect of immediate DC voltage.

Courtesy General Electric Co.

Table 4. Suggested silicon-rectifier replacements for directly heated filament vacuum-tube rectifiers.

<table>
<tr><th rowspan="2">TUBE RECTIFIER</th><th rowspan="2">SYMBOL</th><th colspan="4">TUBE RATING</th><th colspan="8">SILICON RECTIFIER REPLACEMENTS</th></tr>
<tr><th>Max. RMS Voltage</th><th>PK Reverse Voltage</th><th>PK Current Amps</th><th>Max. DC Current Amps Per Anode</th><th>SYMBOL</th><th>Type[1]</th><th>Max. RMS Voltage Transformer Sec.[3]</th><th>PK Reverse Voltage Rating</th><th>Max. Transient Voltage Rating</th><th>Max. DC Current 100° C Ambient</th><th>Thyrector For Secondary</th></tr>
<tr><td>5T4
5U4G
5U4GA
5U4GB
5V3
5V3A
5W4, GT
5Y3
5Y3GA, GT</td><td></td><td>450
450
450
450
425
550
350
350
350</td><td>1550
1550
1550
1550
1400
1550
1400
1400
1400</td><td>.675
.8
.9
1.0
1.2
1.4
.3
.375
.44</td><td>.113
.113
.125
.138
.175
.208
.050
.063
.063</td><td></td><td rowspan="4">1N1693
1N1694
1N1695
1N1696
1N1697
1N560
1N561</td><td rowspan="4">140
180
240
282
325
385
485[2]</td><td rowspan="4">200
300
400
500
600
800
1000[2]</td><td rowspan="4">350
450
600
700
800
960
1200[2]</td><td rowspan="4">.250
.250
.250
.250
.250
.250
.250</td><td rowspan="4">6RS20SP
or
6RS20SC
—5B5
—6B6
—8B8
—9B9
—11B11
—13B13
—16B16</td></tr>
<tr><td>5X4G
5X4GA
5Y4
5Y4GA, GT</td><td>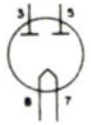</td><td>450
450
350
350</td><td>1550
1550
1400
1400</td><td>.675
.9
.375
.4</td><td>.112
.125
.063
.063</td><td></td></tr>
<tr><td>5Z3
80</td><td></td><td>450
350</td><td>1550
1400</td><td>.675
.400</td><td>.113
.063</td><td>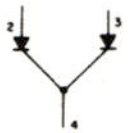</td></tr>
<tr><td>81</td><td></td><td>235</td><td>700</td><td>.5</td><td>.085</td><td></td></tr>
</table>

NOTES:
[1] Any of the silicon rectifiers listed will handle the current. Select on the basis of voltage.
[2] For higher voltages, use rectifiers in series.
[3] The maximum RMS voltage of the transformer secondary has been calculated from the rating of the Thyrector diode (6RS20SP4B4), if used on the 120 volt AC transformer primary. Transient voltages will be suppressed below 175% of the rated peak (√2 RMS) transformer primary and secondary voltages.

Courtesy General Electric Co.

in terms of maximum voltage, peak-inverse voltage, peak current in amperes, and maximum d-c current in amperes for each plate is shown in the tables. The silicon rectifier replacements are direct replacements for the tubes as shown. Where the peak inverse or reverse voltage of the circuit is greater than 1000 volts a single silicon rectifier diode is not recommended for direct substitution. It is possible, however, to use two or more silicon rectifiers connected in series to obtain any desired voltage. Note that in the original circuit in Fig. A the peak-inverse voltage is 920 volts, and because of this, two silicon diodes, type 1N1697, are suggested to replace the rectifier tube. Using this technique it is possible to build up a series of rectifier diodes to almost any desired peak-inverse voltage characteristic of the series.